Singular points
of smooth mappings

C G Gibson

University of Liverpool

Singular points of smooth mappings

Pitman

LONDON · SAN FRANCISCO · MELBOURNE

PITMAN PUBLISHING LIMITED
39 Parker Street, London WC2B 5PB

FEARON PITMAN PUBLISHERS INC.
6 Davis Drive, Belmont, California 94002, USA

Associated Companies
Copp Clark Pitman, Toronto
Pitman Publishing New Zealand Ltd, Wellington
Pitman Publishing Pty Ltd, Melbourne

First published 1979

AMS Subject Classification: 57D45

British Library Cataloguing in Publication Data
Gibson, C G
  Singular points of smooth mappings — (Research notes
  in mathematics; 25).
  1. Mappings (Mathematics) 2. Singularities
  (Mathematics)
   I. Title II. Series
   515   QA360
ISBN 0 273 08410 0

© C G Gibson 1979

All rights reserved. No part of this publication may be reproduced, stored in a retrieval system, or transmitted in any form or by any means, electronic, mechanical, photocopying, recording and/or otherwise without the prior written permission of the publishers. The paperback edition of this book may not be lent, resold, hired out or otherwise disposed of by way of trade in any form of binding or cover other than that in which it is published, without the prior consent of the publishers.

Reproduced and printed by photolithography
in Great Britain at Biddles of Guildford.

TO

DORLE

Contents

Introduction		1
Acknowledgements		7

Chapter I — Smooth Manifolds and Mappings

§1.	A Preliminary Review of Some Calculus	8
§2.	Smooth Manifolds	12
§3.	The Differential of a Smooth Mapping	16
§4.	Vector Fields and Flows	24
§5.	Germs of Smooth Mappings	33

Chapter II — Transversality

§1	The Notion of Transversality	38
§2.	The Basic Transversality Lemma	48
§3.	An Elementary Transversality Theorem	51
§4.	Thom's Transversality Theorem	53
§5.	First Order Singularity Sets	54

Chapter III — Unfoldings : The Finite Dimensional Model

§1.	Groups Acting on Sets	61
§2.	Some Geometry of Jets	62
§3.	Smooth Actions of Lie Groups on Smooth Manifolds	73

| | §4. | Transversal Unfoldings | 81 |
| | §5. | Versal Unfoldings | 89 |

Chapter IV — Singular Points of Smooth Functions

	§1.	Some Basic Geometric Ideas	94
	§2.	The Algebra $\mathcal{E}_n$	99
	§3.	Determinacy of Germs	116
	§4.	Classification of Germs of Codimension ≤ 5	122

Chapter V — Stable Singularities of Smooth Mappings

	§1.	The Basic Ideas	139
	§2.	Contact Equivalence	143
	§3.	Deformations under $\mathcal{K}$-Equivalence	159
	§4.	Classification of Stable Germs	168
	§5.	Higher Order Singularity Sets	174
	§6.	Classifying Germs under $\mathcal{K}$-Equivalence	191
	§7.	Some Examples of Classifying Stable Germs	199
	§8.	Singular Points of Stable Mappings	205

Appendix A — The Theorem of Sard — 215
Appendix B — Semialgebraic Group Actions — 222
Appendix C — Real Algebras — 226
Appendix D — The Borel Lemma — 228
Appendix E — Guide to Further Reading — 232

Index — 237

Introduction

Suppose you take a smooth curved surface X made of some transparent material and imagine it projected downwards onto a plane surface Y by shining a beam of light from above. Think of this as a map $f : X \to Y$ with every point p on the surface X projected down to a point $q = f(p)$ in the plane Y. On the plane you will see the <u>apparent outline</u> of the surface, as it would appear from below. Here are two simple examples.

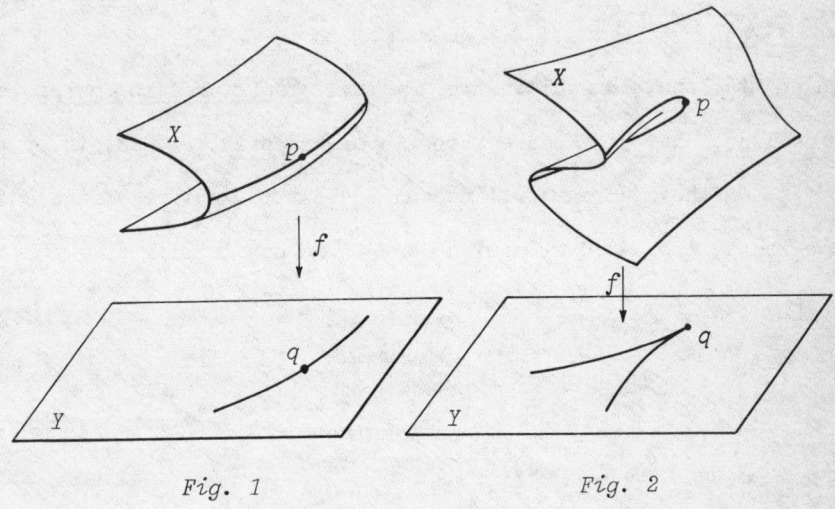

Fig. 1   Fig. 2

It is not hard to see which points on X give rise to the apparent outline: they are precisely the points where the tangent plane to the surface is vertical, the so-called <u>singularities</u> of the mapping $f : X \to Y$. Figure 1 represents the simplest situation one can imagine, with the surface X folding over at p: such points p are called <u>fold points</u>. Figure 2 represents a more complex situation, a curve of fold points on which lies an exceptional point p where two folds meet, a so-called <u>cusp point</u>. A still more complicated situation is provided by Figure 3 where X is a torus, i.e. a doughnut - shaped surface. Here one has again curves of fold points, on which lie

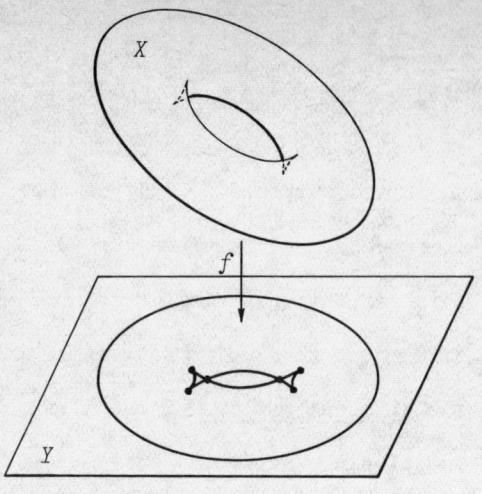

Fig. 3

four cusp points: but in addition we have two <u>simple crossing points</u> on the apparent outline, where the curves cross over properly. Thus, if we take any point q on the apparent outline and look at the nature of the outline very close to q, we can distinguish just three possibilities.

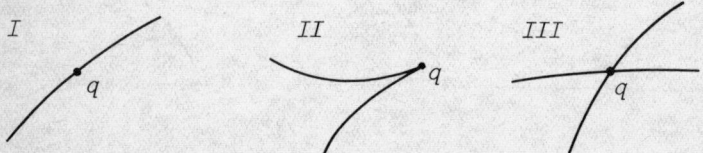

Common sense, and a certain amount of experimentation, will soon convince one that these are the only essential types of behaviour which can arise, in the sense that any other type of behaviour could be eliminated by the slightest change of position of X in space. For instance our torus might be so positioned that the apparent outline was as in Figure 4, with the outline touching itself at some point: but clearly we could just nudge X slightly to get back to the previous situation where only possibilities I, II, III can occur.

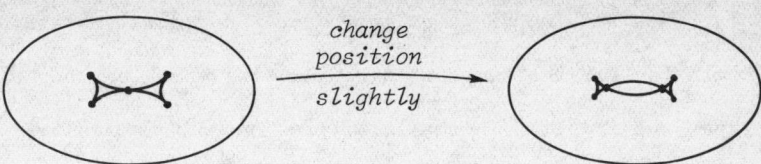

To make the point even clearer, here are three further types of behaviour

which are all inessential, because they could be eliminated by the slightest change in position of X to yield situations exhibiting only fold, cusp and simple crossing points — as follows.

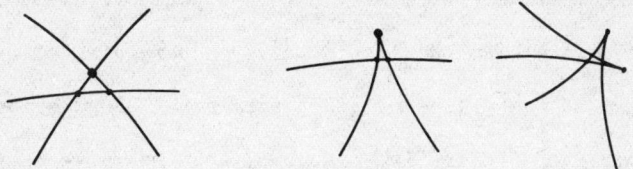

The broad objective of this book is to introduce the reader to the less technical aspects of a mathematical theory of singularities which seeks to make precise the kind of heuristic reasoning just described. The basic objects X, Y are replaced by <u>smooth manifolds</u>, which are natural generalizations to higher dimensions of the familiar notions of curve and surface. And the projection of the surface onto the plane is replaced by an arbitrary <u>smooth mapping</u> $f : X \to Y$. For such mappings we introduce the general notion of <u>singularity</u> and begin to list the simplest singularities which can occur, idealizing each type by a model. We shall, for reasons of simplicity, concern ourselves principally with the <u>local</u> behaviour of the mapping, i.e. its behaviour very close to a single point in the domain: thus in the situation discussed above we interest ourselves solely in the fold and cusp points, and neglect simple crossing points which arise from considering what happens close to two distinct points in the domain.

There is nothing particularly new in the notion of a singularity. Scientists and geometers have recognized them, and appreciated their significance, for a long time now. But no-one seems to have systematically set about studying the singularities of smooth mappings till the pioneering work of Hassler Whitney in the mid 1950's. Around the same time René Thom pointed out the analogy with more finite-dimensional situations and indicated the general lines along which a theory might proceed. So it was in the 1960's that a number of mathematicians, principally John Mather, laid the foundations of a general theory. That was the position in 1967 when Vladimir Arnol'd put together the bits and wrote his now classic survey paper, a model of lucid descriptive writing. It was a time of great promise. Singularity theory itself threw up a number of provocative problems, and the range of possible applications (both within and without mathematics) added to the excitement. Without question, the intervening years have justified that promise, and singularity theory can hold its own as a flourishing area of mathematics.

I feel that the time has come to provide prospective students with readable introductions to the subject. It is my personal conviction that the way to get into any area of mathematics is to concentrate on understanding the simplest situations first, so as to build up some intuitive feeling for what is going on, and to leave the deeper matters till later in life. Singular Points of Smooth Mappings is the result of following this guiding philosophy. I have taken a small number of intuitively appealing ideas and used them to pursue the problem of listing singularity types, one of the goals of the local theory. It is the kind of book which I would expect a postgraduate student in mathematics to read with little difficulty, and I rather hope that others will find it within their compass as well. A guide to further reading has been included to help the reader pursue those matters which interest him most.

A few words are in order concerning the structure of the book. In accordance with the philosophy outlined above smooth manifolds are introduced in Chapter I as subsets of $\mathbb{R}^n$ enjoying certain properties: I think this is the way everyone should meet them. Anyone who wants to get to grips with singularity theory should be familiar with the basic ideas of transversality and of unfolding, so these topics provide the subject matter for Chapters II and III respectively. Here again I have kept to the simplest situations which can arise, imposing restrictions whenever I felt it was possible to suppress undue technicality: in particular, unfoldings have been introduced in a finite-dimensional situation where they are much easier to understand. Singularity theory proper is taken up in Chapter IV with the study of functions; this enables one to make some distance fairly easily, without getting involved in the subtleties associated to general mappings. The result is the derivation of the list of singularities of codimension ≤ 5. In Chapter V the general case of mappings is taken up: it is inevitable that one must quote more and prove less, but I have tried to expose the less technical aspects and give a fairly coherent account of just how one uses the theory to obtain explicit lists of the simplest singularities which turn up. Of course one can pursue the listing process much further than is indicated in this book, but one soon comes up against much deeper matters which lie beyond the scope of an introductory account.

I have not attempted systematically to attribute results to their authors, mainly on the ground that such a practice is out of place in a book at this level. In any case, the material of the first four chapters is now pretty well an established part of the subject. I should say however that the opening sections on differential topology follow closely the exposition given by John Milnor in his excellent little book "Topology, from the Differentiable Viewpoint". The material in the final chapter, basically Mather's classification of stable germs by their local algebras, is not as well-known as it should be. Here I decided to follow the elegant account of Jean Martinet (see the Guide to Further Reading) in which the unfolding idea plays the central role. The key result in this development, namely the characterization of versal unfoldings, turns on a real version of the Weierstrass Preparation Theorem which I do not discuss; I felt it was more important at this level to place proper emphasis on the underlying geometric ideas, and to leave an exposition of the Preparation Theorem to a volume with more ambitious aims.

I decided also to say nothing about the applications of singularity theory, mainly because I feel each area of application is probably worthy of a volume in itself. For instance, Thom's catastrophe theory is already the subject matter of several volumes. Also, the applications within mathematics itself all seem to be at too early a stage to merit writing-up. Here again I hope that my guide to further reading will prove to be useful.

Liverpool
February 1978

Acknowledgements

To Les Lander who drew the pictures, and helped me find enthusiasm at a time when it was all but lost.

To Peter Giblin who undertook the considerable task of correcting the manuscript, and whose suggestions have contributed much to the final form of this book.

And to Ann Garfield who did the typing and produced an excellent job in difficult circumstances.

I Smooth manifolds and mappings

1.   A Preliminary Review of Some Calculus

[In this section it will be understood that U, V, W are <u>open</u> sets in $\mathbb{R}^n$, $\mathbb{R}^p$, $\mathbb{R}^q$ respectively.]

Let $f : U \to V$ be a mapping with components $f_1, \ldots, f_p$. We call f <u>smooth</u> when all the partial derivatives

$$\frac{\partial^\alpha f}{\partial x_1^{\alpha_1} \ldots \partial x_n^{\alpha_n}} \qquad (\alpha = \alpha_1 + \ldots + \alpha_n)$$

exist and are continuous in U. For the purposes of this book it will suffice to observe that if $f_1, \ldots, f_p$ are all given by polynomials in $x_1, \ldots, x_n$ then f will be smooth. And we call f a <u>diffeomorphism</u> of U onto V when it is a bijection, and both f, f^{-1} are smooth.

Now suppose $f : U \to V$ is smooth. For every point $a \in U$ there is a linear mapping $D_a f : \mathbb{R}^n \to \mathbb{R}^p$ called the <u>differential of f at a</u>: it is precisely the linear mapping whose matrix (relative to the standard bases) is the so-called Jacobian

$$\begin{pmatrix} \frac{\partial f_1}{\partial x_1}(a) & \cdots & \frac{\partial f_1}{\partial x_n}(a) \\ \vdots & & \vdots \\ \frac{\partial f_p}{\partial x_1}(a) & \cdots & \frac{\partial f_p}{\partial x_n}(a) \end{pmatrix}.$$

Take, for instance, the case when $f : U \to V$ is the restriction of a linear map $F : \mathbb{R}^n \to \mathbb{R}^p$; then f is certainly smooth, and explicit computation of the Jacobian will verify that $D_a f = F$ at every point $a \in U$. A particular case is provided when $U \subseteq V$ and $f : U \to V$ is the inclusion mapping; in this case $D_a f = 1$ at every point $a \in U$, with 1 the identity mapping on $\mathbb{R}^n$.

One of the basic facts we shall need to know about the differential is the Chain Rule, which we express as follows. If we have a commuting triangle of smooth maps

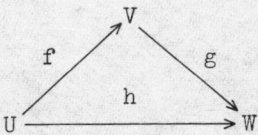

And a point $a \in U$ with $f(a) = b$, then the corresponding triangle of differentials commutes.

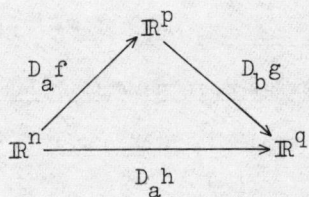

This provides us with a simple proof of the following fact.

(1.1)   If $f : U \to V$ is a diffeomorphism, then at each point $a \in U$ the differential $D_a f : \mathbb{R}^n \to \mathbb{R}^p$ is invertible, so that necessarily $n = p$.

Proof   Indeed we shall have the commuting triangle of smooth mappings

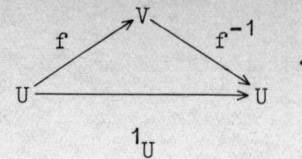

Now let $a \in U$, and put $b = f(a)$. By the Chain Rule we have a commuting diagram of differentials

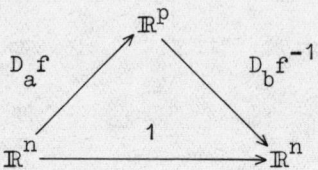

which entails that $D_a f$ is invertible, with inverse $D_b f^{-1}$. □

The reader should be reminded that the direct converse of (1.1) does not hold. Nevertheless, there is a partial converse, the Inverse Function Theorem, which provides one of the really basic theorems of the calculus.

(1.2)   <u>If $f : U \to V$ is smooth, and $a \in U$ is such that $D_a f : \mathbb{R}^n \to \mathbb{R}^n$ is invertible, then there exist open neighbourhoods U', V' of a, $f(a)$ respectively such that f maps U' diffeomorphically onto V'.</u>

The next two propositions are important consequences of the Inverse Function Theorem and provide the keys to several useful results.

(1.3)   <u>Let $f : U \to \mathbb{R}^p$ be a smooth mapping with $f(0) = 0$ and $D_0 f$ of rank p (so necessarily $n \geq p$): then there exists a diffeomorphism h of some neighbourhood of 0 in $\mathbb{R}^n$ onto another such that $h(0) = 0$ and</u>

$$f \circ h^{-1}(x_1, \ldots, x_n) = (x_1, \ldots, x_p).$$

Proof   The Jacobian matrix of f will have rank n: we can suppose it is the leading $p \times p$ submatrix which is invertible.   Define a smooth mapping F from a neighbourhood of 0 in $\mathbb{R}^n$ onto another by

$$F(x_1, \ldots, x_n) = \left(f_1(x), \ldots, f_p(x), x_{p+1}, \ldots, x_n\right)$$

where $f_1, \ldots, f_p$ are the components of f.   Computation verifies that $D_0 F$ is invertible.   By the Inverse Function Theorem F has a restriction h which is a diffeomorphism of a neighbourhood of 0 in $\mathbb{R}^n$ onto another, and automatically satisfies $h(0) = 0$.   To see that h is the required diffeomorphism take $\pi : \mathbb{R}^n \to \mathbb{R}^p$ to be the projection defined by $\pi(x_1, \ldots, x_n) = (x_1, \ldots, x_p)$ and observe that

$$f \circ h^{-1}(x_1, \ldots, x_n) = \pi \circ F \circ h^{-1}(x_1, \ldots, x_n) = (x_1, \ldots, x_p). \quad \square$$

(1.4)   Let $f : U \to \mathbb{R}^p$ be a smooth mapping with $f(0) = 0$ and $D_0 f$ of rank n (so necessarily $n \leq p$): then there exists a diffeomorphism k of some neighbourhood of 0 in $\mathbb{R}^p$ onto another such that $k(0) = 0$ and

$$k \circ f(x_1, \ldots, x_n) = (x_1, \ldots, x_n, 0, \ldots, 0).$$

Proof   The Jacobian matrix of f will have rank n: we can suppose it is the leading $n \times n$ submatrix which is invertible.   Define a smooth mapping F from a neighbourhood of 0 in $\mathbb{R}^p$ onto another by

$$F(x_1, \ldots, x_p) = f(x_1, \ldots, x_n) + (0, \ldots, 0, x_{n+1}, \ldots, x_p).$$

Computation will verify that $D_0 F$ is invertible.   By the Inverse Function Theorem F has a restriction which is a diffeomorphism of a neighbourhood of 0 in $\mathbb{R}^p$ onto another.   Denote the inverse of this restriction by k: clearly it is the required diffeomorphism as

$$f(x_1, \ldots, x_n) = F(x_1, \ldots, x_n, 0, \ldots, 0).\qquad \square$$

That does not conclude our review of calculus, for we shall also require standard results concerning the existence of solutions for ordinary differential equations. However we shall postpone that discussion till §4 when we shall have available the language necessary to state results in a succinct geometric form.

§2. Smooth Manifolds

It is an easy matter to extend the basic concept of "smoothness" to mappings between arbitrary subsets X, Y of $\mathbb{R}^j, \mathbb{R}^k$. We call $f : X \to Y$ <u>smooth</u> when for any point $x \in X$ there exists an open neighbourhood U of x in $\mathbb{R}^j$ (depending on x) and a smooth mapping $F : U \to \mathbb{R}^k$ with $f = F$ on $U \cap X$. And, by analogy with §1, we call $f : X \to Y$ a <u>diffeomorphism</u> when f is bijective, and both f, f^{-1} are smooth: in that case X, Y are said to be <u>diffeomorphic</u>.

However, we do not wish to consider completely arbitrary subsets of Euclidean spaces: instead we shall isolate a particularly useful class of subsets. First we need some definitions. An <u>n-dimensional parametrization</u> of a set $X \subseteq \mathbb{R}^j$ is a smooth mapping $\phi : V \to \mathbb{R}^j$ with V an open set in $\mathbb{R}^n$ for which $\phi(V) = X$ and the mapping $\phi : V \to \phi(V)$ is a diffeomorphism. Suppose now that $N \subseteq \mathbb{R}^k$ and that $x \in N$: a <u>relatively open neighbourhood</u> of x in N is a set of the form $U \cap N$ with $U \subseteq \mathbb{R}^k$ an open set for which $x \in U$. Now we are ready to introduce the main idea. We call $N \subseteq \mathbb{R}^k$ a <u>smooth manifold of dimension n</u> when every point $x \in N$ has a relatively open neighbourhood $U \cap N$ admitting an n-dimensional parametrization $\phi : V \to \mathbb{R}^k$: some authors term the inverse $\phi^{-1} : U \cap N \to V$ a <u>chart at x</u>, and call its components <u>local co-ordinates at x</u>. We shall adhere to the convention of writing N^n to indicate that N has dimension n; and sometimes we shall write dim N for the dimension of N.

<u>Example 1</u>   Any open set $N \subseteq \mathbb{R}^k$ is a smooth manifold of dimension k. (and in particular $\mathbb{R}^k$ is one). Here, given $x \in N$ one takes $U = \mathbb{R}^k$, and then the inclusion $N \to \mathbb{R}^k$ is a k-dimensional parametrization of $U \cap N$.

<u>Example 2</u>   Any vector subspace $N \subseteq \mathbb{R}^k$ of dimension n (as a vector space) is a smooth manifold of dimension n. Given $x \in N$ we take $U = \mathbb{R}^k$ and then any linear mapping $\mathbb{R}^n \to \mathbb{R}^k$ with image N will be an n-dimensional parametrization of $U \cap N$.

In both these examples the whole set N admits a parametrization. The next example is more subtle in that the set in question is compact, so cannot possibly admit a parametrization; however it is still a smooth manifold.

<u>Example 3</u>   Let $S^n \subseteq \mathbb{R}^{n+1}$ be the n-sphere, i.e. the set of all points $(x_0, x_1, \ldots, x_n)$ with $x_0^2 + \ldots + x_n^2 = 1$. I claim that S^n is a smooth manifold of dimension n. For convenience we take the <u>north pole</u> and the <u>south pole</u> of S^n to be the points

$$N = (1, 0, \ldots, 0) \qquad S = (-1, 0, \ldots, 0)$$

and we take

$$\phi_N : S^n - \{N\} \to \mathbb{R}^n$$

$$\phi_S : S^n - \{S\} \to \mathbb{R}^n$$

to be respectively stereographic projection from the north, south pole, i.e.

the mappings defined by

$$\phi_N(x_0, \ldots, x_n) = \left(\frac{x_1}{1 - x_0}, \ldots, \frac{x_n}{1 - x_0}\right)$$

$$\phi_S(x_0, \ldots, x_n) = \left(\frac{x_1}{1 + x_0}, \ldots, \frac{x_n}{1 + x_0}\right).$$

Since the domains of ϕ_N, ϕ_S cover S^n it will suffice to show that both these mappings are diffeomorphisms. They are certainly smooth, being the restrictions of the smooth mappings defined by the same formulae with domains the open sets defined by $x_0 \neq 1$, $x_0 \neq -1$ respectively. A minor computation will verify that ϕ_N, ϕ_S are bijections with inverses defined by

$$\phi_N^{-1}(y_1, \ldots, y_n) = \frac{1}{1 + |y|^2}\left(|y|^2 - 1, 2y_1, \ldots, 2y_n\right)$$

$$\phi_S^{-1}(y_1, \ldots, y_n) = \frac{1}{1 + |y|^2}\left(1 - |y|^2, 2y_1, \ldots, 2y_n\right)$$

where

$$y = (y_1, \ldots, y_n) \qquad |y|^2 = y_1^2 + \ldots + y_n^2.$$

And clearly the inverses are smooth mappings as well. In the case when $n = 2$ one pictures ϕ_N, ϕ_S like this.

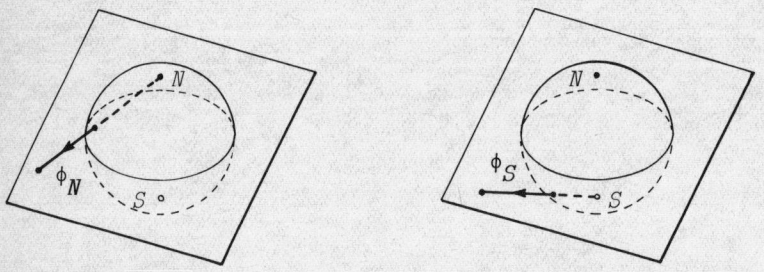

The reader will no doubt note, with increasing apprehension, that even for such a simple example as S^n a fair amount of work is involved in verifying that it is indeed a smooth manifold. However in Chapter II we shall establish a proposition which provides a very simple way of showing that quite complicated sets are smooth manifolds. For the time being we shall stick to our rather short list, but observe that one can increase the stock of examples using

(2.1) <u>Let $M \subseteq \mathbb{R}^j$, $N \subseteq \mathbb{R}^k$ be smooth manifolds of dimensions m, n respectively: then $M \times N \subseteq \mathbb{R}^{j+k}$ is a smooth manifold of dimension</u> $(m + n)$.

The proof of (2.1) is sufficiently obvious to justify its omission. By way of illustration the torus $S^1 \times S^1 \subseteq \mathbb{R}^4$ is a smooth manifold of dimension 2.

Let $N^n \subseteq \mathbb{R}^k$ be a smooth manifold. A <u>smooth submanifold</u> of N is a smooth manifold $M^m \subseteq \mathbb{R}^k$ with $M \subseteq N$. One calls $n - m$ the <u>codimension</u> of M in N. The next result tells us that, locally, a smooth submanifold of a smooth manifold looks like a subspace of a vector space.

(2.2) <u>Let M^m be a smooth submanifold of a smooth manifold $N^n \subseteq \mathbb{R}^k$: any point $x \in M$ has a relatively open neighbourhood in N having a parametrization $\phi : V \to \mathbb{R}^k$ for which $\phi(0) = x$ and $\phi(\mathbb{R}^m \times 0) = \phi(V) \cap M$.</u>

Proof Let μ, ν be parametrizations of dimensions m, n of relatively open neighbourhoods of x in M, N respectively. Clearly, one can suppose $\mu(0) = x$, $\nu(0) = x$. Now $f = \nu^{-1} \circ \mu$ is a smooth mapping of an open subset of $\mathbb{R}^m$ into $\mathbb{R}^n$ with $f(0) = 0$ whose differential at 0 has rank m. (Note that therefore $m \leq n$.) By (1.4) there exists a diffeomorphism k of some open neighbourhood of 0 in $\mathbb{R}^n$ onto another for which

$$k_0 f(x_1, \ldots, x_m) = (x_1, \ldots, x_m, 0, \ldots, 0)$$

and then $\phi = \nu_0 k^{-1}$ has the desired properties. □

In differential topology (the study of smooth manifolds and smooth mappings between them) one regards two smooth manifolds N_1, N_2 as being the "same" when they are diffeomorphic. Likewise we regard two smooth mappings $f_1 : N_1 \to P_1$, $f_2 : N_2 \to P_2$ as being the "same", formally one calls them <u>equivalent,</u> when there exist diffeomorphisms h, k for which the following diagram commutes.

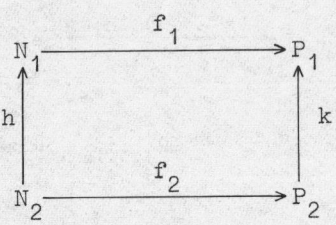

§3. The Differential of a Smooth Mapping

Our objective in this section is to introduce the concept of "differential" for smooth mappings with domain a smooth manifold, rather than just an open set in some Euclidean space. The following definition provides a necessary preliminary. Let $N^n \subseteq \mathbb{R}^k$ be a smooth manifold, let $x \in N$ and let $\phi : U \to \mathbb{R}^k$ be a parametrization of a relatively open neighbourhood of x in N with $\phi(u) = x$, say. We define the <u>tangent space</u> $T_x N$ at x to N to be the image of the differential $D_u \phi : \mathbb{R}^n \to \mathbb{R}^k$. One pictures $T_x N$ as the vector subspace of $\mathbb{R}^k$ parallel to the affine subspace of $\mathbb{R}^k$ through x which best approximates N close to x.

Of course, for the above definition to make sense we must show that it is independent of the choice of parametrization. So let $\psi : V \to \mathbb{R}^k$ be another parametrization of a relatively open neighbourhood of x in N with $\psi(v) = x$, say. Then $\chi = \psi^{-1} \circ \phi$ will map an open neighbourhood U_1 of u onto an open neighbourhood V_1 of V, and we shall have a commuting diagram of smooth mappings

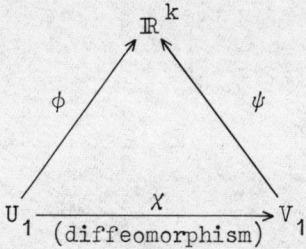

giving rise to a commuting diagram of differentials

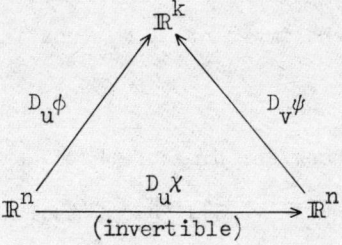

from which it is immediate that $D_u\phi$, $D_v\psi$ have the same image, as was required.

(3.1) Let $N \subseteq \mathbb{R}^k$ be a smooth manifold, and let $x \in N$: then the tangent space $T_x N$ is a vector subspace of $\mathbb{R}^k$ of the same dimension n as N.

Proof   We keep to the notation of the above discussion, i.e. $\phi : U \to \mathbb{R}^k$ is a parametrization of a relatively open neighbourhood of x in N, with $\phi(u) = x$. Now $\phi^{-1} : \phi(u) \to U$ is smooth, so (by definition) there is an open set $W \subseteq \mathbb{R}^n$ with $\phi^{-1} = \Phi$ on $W \cap \phi(U)$. Taking U' to be the inverse image under ϕ of $W \cap \phi(u)$ we obtain a commuting diagram of smooth mappings

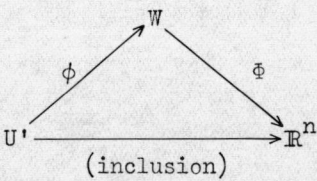

hence a commuting diagram of differentials

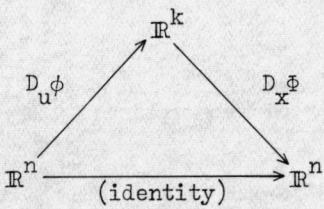

It is immediate that $T_x N$, the image of $D_u \phi$, has the required dimension n.   □

The reader will readily check for himself that the tangent space at a point to an open set in $\mathbb{R}^k$ is precisely $\mathbb{R}^k$: also, that the tangent space at a point to a vector subspace V of $\mathbb{R}^k$ is precisely V. A harder exercise is to show that the tangent space at a point x to the n-sphere $S^n \subseteq \mathbb{R}^{n+1}$ is precisely the subspace perpendicular to x. However for such examples as this (where the smooth manifold is defined by equations) we shall establish much simpler ways of computing tangent spaces in Chapter II.

The importance of the tangent space is that it allows one to introduce the differential at a point of a smooth mapping defined on a smooth manifold, rather than just on an open set. The definition will require a preliminary observation. Let $M^m \subseteq \mathbb{R}^p$, $N^n \subseteq \mathbb{R}^q$ be two smooth manifolds, let $f : M \to N$ be a smooth mapping, let $x \in M$, and put $y = f(x)$. The fact that f is smooth requires that there is an open neighbourhood W of x in $\mathbb{R}^p$, and a smooth mapping $F : W \to \mathbb{R}^q$ with $f = F$ on $W \cap M$. Of course the differential of F at x is a linear mapping $D_x F : \mathbb{R}^p \to \mathbb{R}^q$. We claim that $D_x F$ necessarily maps the tangent space $T_x M$ into the tangent space $T_y N$. To prove this let $\phi : U \to \mathbb{R}^p$, $\psi : V \to \mathbb{R}^q$ be parametrizations of relatively open neighbourhoods of x, y in M, N respectively with $\phi(u) = x$, $\psi(v) = y$, say. By taking U to be sufficiently small we can suppose that $\phi(U) \subseteq W$, and that $f \circ \phi(U) \subseteq \psi(V)$, so $g : U \to V$ given by $g = \psi^{-1} \circ f \circ \phi$ is a well-defined smooth mapping. We now have a commuting diagram of smooth mappings

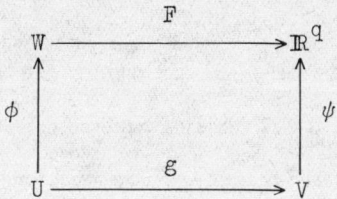

yielding a commuting diagram of differentials

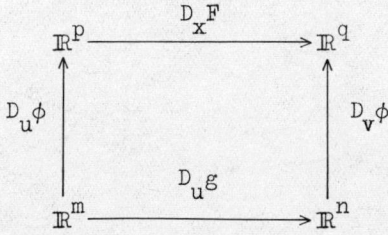

It is immediate that $D_x F$ maps the image of $D_u \phi$ into the image of $D_v \psi$, i.e. the tangent space T_xM into the tangent space T_yN, as was claimed. The restriction of the linear mapping $D_xF : \mathbb{R}^p \to \mathbb{R}^q$ to T_xM we write $T_xf : T_xM \to T_yN$ and call the <u>differential of</u> f <u>at</u> x. One pictures T_xf as the best linear approximation to f at x: the illustration below is drawn for the case $x = 0$ in $\mathbb{R}^p$, and $y = 0$ in $\mathbb{R}^q$.

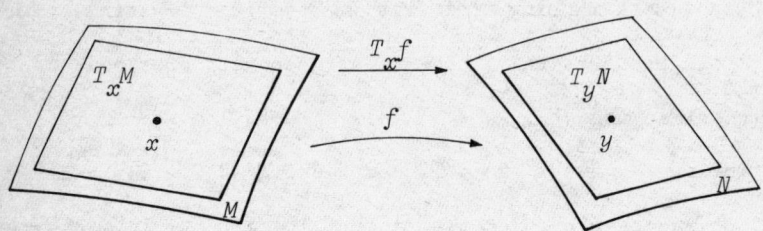

It follows from the very construction of T_xf that it does not depend on the choice of F. Notice also that in the particular case when M, N are open sets in $\mathbb{R}^p$, $\mathbb{R}^q$ the differential T_xf is the the differential D_xf as previously defined. It is an easy matter to derive the basic properties of this more general notion of differential from those mentioned in §1. We shall mention two of these, leaving the proofs to the reader.

Take, for instance, the situation when M is a smooth submanifold of a smooth manifold N, and $f : M \to N$ is the inclusion mapping. At any point $x \in M$ the tangent space T_xM is a subspace of T_xN, and the differential $T_xf : T_xM \to T_xN$ is the inclusion. One pictures the situation something like this.

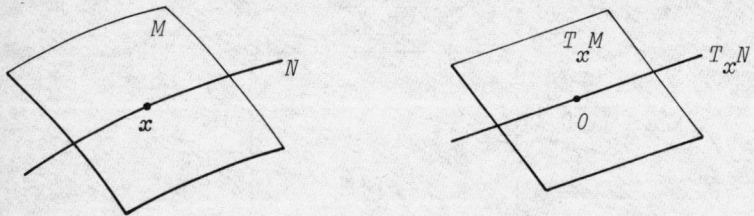

Likewise, we have the Chain Rule, which we express as follows. Suppose A, B, C are smooth manifolds, and we have a commuting triangle of smooth mappings

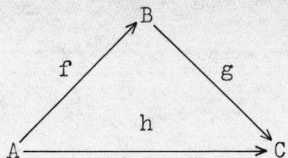

and a point $a \in A$ with $f(a) = b$, $g(b) = c$ then the corresponding triangle of differentials commutes

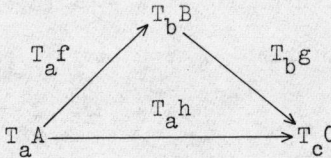

And as in §1 it follows from these two basic properties that at any point the differential of a diffeomorphism is an isomorphism of vector spaces, so that in particular domain and target have the same dimension. Our next result provides us with further examples of smooth manifolds. First a definition. We define the <u>graph</u> of a mapping $f : N \to P$ to be the set

$$\text{graph } f = \{(x, y) \in N \times P : y = f(x)\}.$$

(3.2) <u>Let $f : N \to P$ be a smooth mapping between smooth manifolds: then graph f is a smooth submanifold of $N \times P$. Also the tangent space to graph f at any point $(x, f(x))$ is precisely graph $T_x f$.</u>

<u>Proof</u> $F : N \to$ graph f given by $F(x) = (x, f(x))$ is a smooth bijection, indeed a diffeomorphism since its inverse F^{-1} is the restriction to graph f of the smooth projection $N \times P \to N$. It follows immediately that

graph f is a smooth manifold. Now $T_x F$ is the mapping $(1, T_x f)$ and the tangent space to graph f will be its image, i.e. graph $T_x f$. □

At this point is is convenient to augment (2.1) by the following proposition, determining the tangent space to a product; the proof is left to the reader.

(3.3) <u>Let</u> M, N <u>be smooth manifolds.</u> <u>The tangent space at a point</u> (x, y) <u>to the product</u> M × N <u>is the product</u> $T_x M \times T_y N$.

So far we have talked only about the differential $T_x f$ of a smooth mapping f <u>at a point</u> x. We shall conclude this section by mentioning an elegant globalisation of this notion which is essential to any serious study of differential topology. The idea is to glue together all the linear mappings $T_x f$ to obtain a single mapping Tf. Formally, we proceed as follows. Let $N \subseteq \mathbb{R}^k$ be a smooth manifold. We define the <u>tangent bundle space</u> to N to be the set of all possible tangent vectors to N, i.e.

$$TN = \{(x, v) \in N \times \mathbb{R}^k : v \in T_x N\}.$$

(3.4) <u>Let</u> $N \subseteq \mathbb{R}^k$ <u>be a smooth manifold of dimension</u> n: <u>the tangent bundle space</u> $TN \subseteq \mathbb{R}^{2k}$ <u>is a smooth manifold of dimension</u> 2n.

<u>Proof</u> Let $f : U \to \mathbb{R}^k$ be an n-dimensional parametrization of a relatively open set in N. The proposition follows from the observation that $F : U \times \mathbb{R}^n \to \mathbb{R}^k \times \mathbb{R}^k$ defined by $(u, x) \to \bigl(f(u), D_u f(x)\bigr)$ is a 2n-dimensional parametrization of a relatively open set in TN. □

Now let $f : N \to P$ be a smooth mapping with N, P smooth manifolds. We define the <u>tangent mapping</u> Tf : TN → TP by the formula

$$Tf(x, v) = \bigl(f(x), T_x f(v)\bigr).$$

(3.5)   Let $f : N \to P$ be a smooth mapping: the tangent mapping $Tf : TN \to TP$ is likewise smooth.

Proof   Let $N \subseteq \mathbb{R}^s$, $P \subseteq \mathbb{R}^t$. Choose a point $(x_0, v_0) \in TN$ and an open set $U \subseteq \mathbb{R}^s$ with $x_0 \in U$ for which there exists a smooth mapping $F : U \to \mathbb{R}^t$ with $F = f$ on $U \cap N$. (This is possible since f is smooth.) Observe that the tangent mapping of F is the map $TF : U \times \mathbb{R}^s \to \mathbb{R}^t \times \mathbb{R}^t$ given by $(x, v) \to \bigl(F(x), D_x F(v)\bigr)$ and that this mapping is smooth. But $U \times \mathbb{R}^s \subseteq \mathbb{R}^s \times \mathbb{R}^s$ is open and contains (x_0, v_0): also we have $TF = Tf$ on $(U \times \mathbb{R}^s) \cap TN$, going back to the definition of Tf. It follows that Tf is smooth.   □

Notice that if M is a smooth submanifold of a smooth manifold N then the tangent mapping of the inclusion $M \to N$ will be the inclusion $TM \to TN$. One of the advantages of the tangent mapping is that it allows a simple and elegant formulation of the Chain Rule, namely that a commuting diagram of smooth mappings

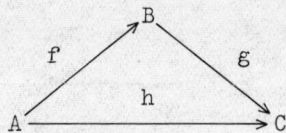

gives rise to a commuting diagram of tangent mappings

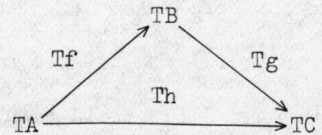

otherwise expressed,

$$T(g \circ f) = Tg \circ Tf.$$

Notice also that given a smooth manifold N there is a smooth projection $\pi_N : TN \to N$ given by the formula $(x, v) \to x$: this mapping is the <u>tangent bundle</u> of N. It follows that given any smooth mapping $f : N \to P$ between smooth manifolds we have a commuting diagram of smooth mappings.

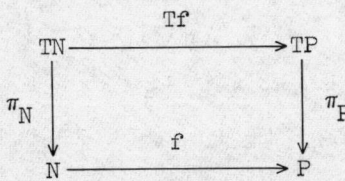

§4. Vector Fields and Flows

Let N^n be a smooth manifold with tangent bundle $\pi : TN \to N$. By a <u>smooth vector field</u> on N we mean a smooth section of π, i.e. a smooth mapping $\xi : N \to TN$ such that $\pi \circ \xi = 1_N$. In other words ξ assigns to each point $x \in N$ a tangent vector $\xi(x)$ at x.

For the purpose of illustration take the special case when $N \subseteq \mathbb{R}^n$ is an open set. In that case TN is $N \times \mathbb{R}^n$ and $\pi : N \times \mathbb{R}^n \to N$ is projection on the first factor, so that a smooth vector field is a smooth mapping $\xi : N \to N \times \mathbb{R}^n$ of the form $x \to \bigl(x, \theta(x)\bigr)$ with $\theta : N \to \mathbb{R}^n$ a smooth mapping. Suppose $\theta_1, \ldots, \theta_n$ are the components of θ: then our vector field does no more than associate with each point $x \in N$ the vector $\bigl(\theta_1(x), \ldots, \theta_n(x)\bigr)$ in $\mathbb{R}^n$. Thus, for instance, in the case when $n = 2$

we can sketch a vector field by the simple device of drawing at each point $x \in N$ an arrow starting at that point, and in the direction given by the vector $\bigl(\theta_1(x), \theta_2(x)\bigr)$. Here are some examples, which the reader is advised to check for himself: indeed a good exercise is to invent simple formulae for θ and sketch the corresponding vector fields. In all these examples we have taken $N = \mathbb{R}^2$.

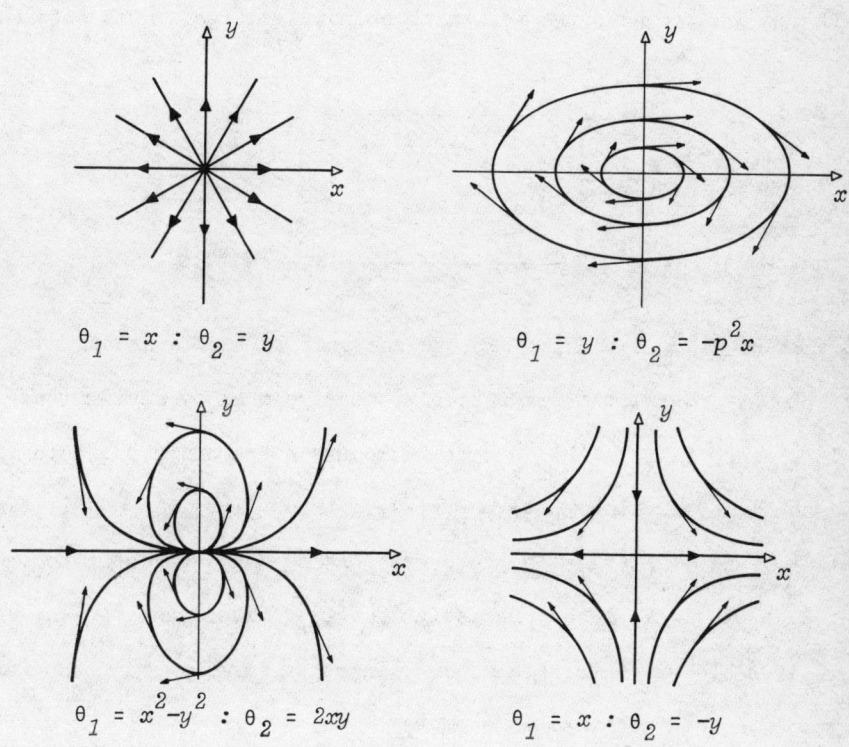

$\theta_1 = x : \theta_2 = y$

$\theta_1 = y : \theta_2 = -p^2 x$

$\theta_1 = x^2 - y^2 : \theta_2 = 2xy$

$\theta_1 = x : \theta_2 = -y$

These pictures have a certain didactic value. They suggest that if you start at a point and follow the arrows you will move along a curve, a "flow line" if you like. We shall make this precise. Let N^n be a smooth manifold. By a <u>smooth curve</u> in N we mean a smooth mapping $f : I \to N$ with I an open interval of positive length. For convenience we shall suppose that 0 is the mid-point of I. And we say that the curve starts at $x_0 = f(0)$.

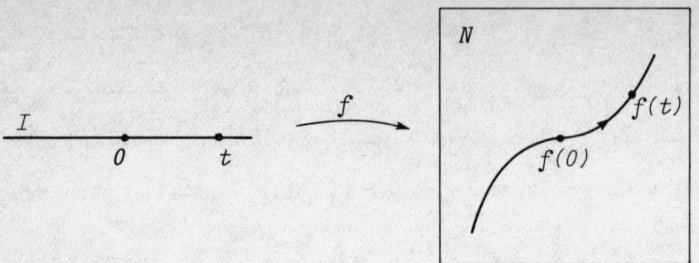

Recall now that we have the following commuting diagram of smooth mappings.

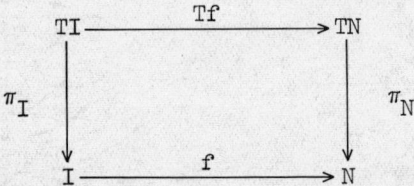

Here, of course, $TI = I \times \mathbb{R}$. Suppose now that we have a smooth vector field ξ on N. We wish to make precise what we mean by saying that f is a "flow-line" for ξ. To this end we introduce a "canonical" smooth vector field i on I by writing $i(t) = (t, 1)$. Thus $i(t)$ is a unit tangent vector at t, and $Tf(i(t))$ is the corresponding tangent vector to the curve at $f(t)$. We want this to be precisely $\xi(f(t))$, i.e. the tangent vector at $f(t)$ given by our vector field ξ. In other words we want the following diagram of smooth mappings to commute

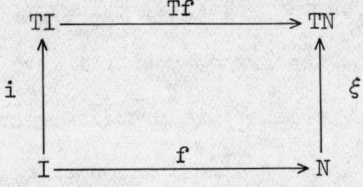

Thus we are led to define a <u>flow-line</u> for ξ to be a smooth curve $f : I \to N$ for which the above diagram commutes. The composite mapping $Tf \circ i$ is

26

usually written f': with that notation the condition for f to be a flow-line is that

$$f'(t) = \xi(f(t)). \quad *$$

To make this as concrete as possible it might help to go back to the case when N is an open set in $\mathbb{R}^n$, and the vector field is given by n smooth functions $\theta_1, \ldots, \theta_n$ on N. A smooth curve f in N will have n smooth functions $f_1, \ldots, f_n$ as its components. And the condition * for f to be a flow-line of the vector field is, written out in full, the following system of n simultaneous equations

$$f'_1(t) = \theta_1(f_1(t), \ldots, f_n(t))$$
$$\vdots$$
$$f'_n(t) = \theta_n(f_1(t), \ldots, f_n(t)).$$

Thus the problem of constructing a flow-line f for ξ is that of solving simultaneously the above n differential equations for $f_1, \ldots, f_n$. And the condition for the flow-line to start at the point $(x_1, \ldots, x_n)$ in N is that the solution should satisfy the "initial conditions"

$$f'_1(0) = x_1 : \ldots : f'_n(0) = x_n.$$

<u>Example 1</u>   Take the smooth vector field on $\mathbb{R}^2$ defined by $\theta_1 = y$, $\theta_2 = -p^2 x$ with $p \neq 0$ a real number. (We sketched this one above.) To find flow-lines we have to solve the simultaneous differential equations

$$f'_1 = f_2 : f'_2 = -p^2 f_1.$$

These yield the harmonic equation

27

$$f_1'' + p^2 f_1 = 0$$

having solutions

$$f_1(t) = c \sin(pt + \alpha) \quad : \quad f_2(t) = p^2 c \cos(pt + \alpha)$$

with $c, \alpha \in \mathbb{R}$. Notice that $p^2 f_1^2 + f_2^2 = p^2 c^2$, so the flow-lines are the ellipses $p^2 x^2 + y^2 = p^2 c^2$. (Compare with the sketch given above.)

The most basic fact which it is necessary to know about flow-lines is what we shall term the Local Existence and Uniqueness Theorem.

(4.1) Let ξ <u>be a smooth vector field on a smooth manifold</u> N, <u>and let</u> $x_0 \in N$. <u>The following assertions hold.</u>

(i) <u>There exists a flow-line for</u> ξ <u>which starts at</u> x_0. (Existence)

(ii) <u>Any two flow-lines for</u> ξ <u>which start at</u> x_0 <u>agree on some neighbourhood of</u> 0 <u>in</u> $\mathbb{R}^n$. (Uniqueness)

The reader will readily check that it suffices to prove the result when N is an open subset of $\mathbb{R}^n$. And in that case the result is just the Local Existence and Uniqueness Theorem for ordinary differential equations, which we assume him to be familiar with from calculus.

In fact we shall need to know rather more than the mere existence of a single flow-line through a point: we need to know that we can simultaneously parametrize all the flow-lines through points near to a given one. This idea is captured by the following formal definition. Let ξ be a smooth vector field on a smooth manifold N, and let x_0 be a point in N. By a <u>smooth local flow</u> for ξ at x_0 we mean a smooth mapping $F : U \times I \to N$ (where U is an open neighbourhood of x_0 in N, and I is an open interval) with the property that for any point $x \in U$ the mapping $f_x : I \to N$ defined by

$f_x(t) = F(x, t)$ is a flow-line for ξ which starts at x. One pictures the situation something like this, with

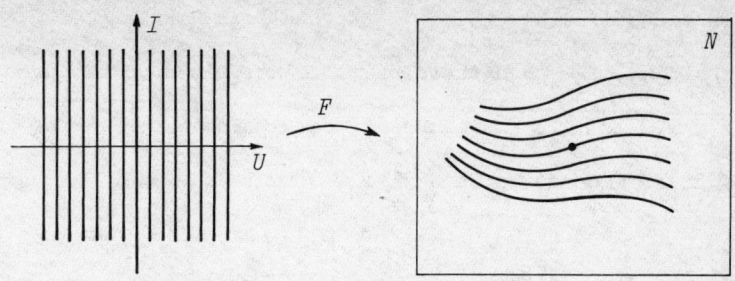

the vertical lines in the left-hand picture mapping to the flow-lines in the right-hand picture. The fact which we need to know is the following result, which again we assume the reader to be familiar with from calculus.

(4.2) Let ξ be a smooth vector field on a smooth manifold N, and let $x_0 \in N$: there exists a smooth local flow for ξ at x_0.

The next thing we need to know is that we can "straighten-out" a local flow. However, for this we need an additional proviso, embodied in the following definition. A critical point for a smooth vector field ξ on a smooth manifold N is a point $x \in N$ for which $\xi(x) = 0$. Notice that at such a point the flow-line is constant, i.e. its image is a point. (Indeed a constant curve through x is certainly a flow-line, and the Local Existence and Uniqueness Theorem tells us that it is the only one.)

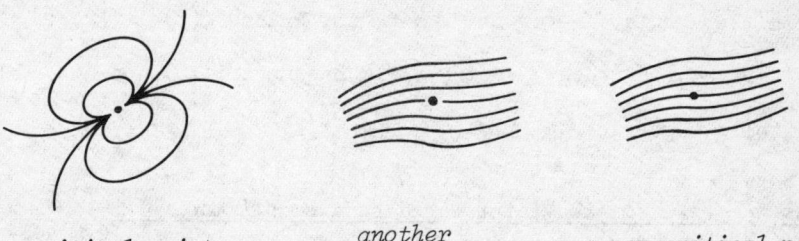

a critical point   *another critical point*   *a non-critical point*

We formalise the idea of "straightening out" a vector field as follows. Suppose that x_1, x_2 are points in smooth manifolds N_1, N_2 on which are defined smooth vector fields ξ_1, ξ_2. We shall say that ξ_1, ξ_2 are <u>smoothly equivalent</u> at x_1, x_2 when there exist relatively open neighbourhoods U_1, U_2 of x_1, x_2 in N_1, N_2 and a diffeomorphism $\Phi : U_1 \to U_2$ for which the following diagram commutes, and $\Phi(x_1) = x_2$.

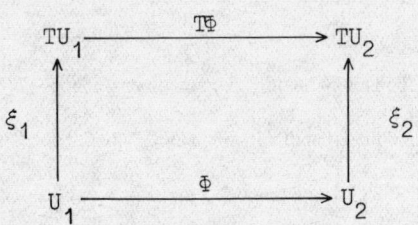

Here of course we are abusing notation by allowing ξ_1, ξ_2 also to denote their restrictions to U_1, U_2 respectively. One pictures the situation something like this with Φ mapping the flow-lines for ξ_1 to those for ξ_2.

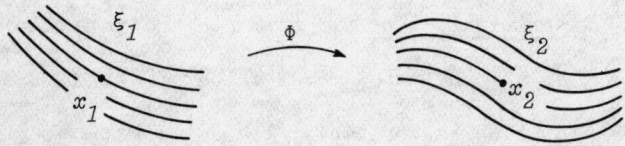

Now let us call a smooth vector field ξ on a smooth manifold N <u>constant</u> when there exists a vector $v \neq 0$ for which $\xi(x) = (x, v)$ for all $x \in N$. Of course, such a vector field has no critical points. We contend the following.

(4.3) Let ξ <u>be a smooth vector field on a smooth manifold</u> N <u>and let</u> $x_0 \in N$ <u>be a non-critical point. There exists a constant vector field on</u> N <u>smoothly equivalent to</u> ξ <u>at</u> x_0.

The picture for this result is the same as that above save that one of the vector fields is constant.

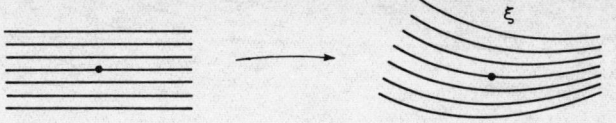

Proof   Clearly, it is no restriction to suppose that N is an open neighbourhood of $x_0 = 0$ in $\mathbb{R}^n$. By (4.2) there exists a smooth local flow $F : U \times I \to \mathbb{R}^n$ with U an open neighbourhood of I in $\mathbb{R}^n$, and I an open interval containing 0. Let $v = \xi(0)$: then $v \neq 0$ as 0 is a non-critical point. I claim that ξ is smoothly equivalent at 0 to the constant vector field on $\mathbb{R}^n$ given by v. For this we have to construct a diffeomorphism Φ of some neighbourhood of 0 in $\mathbb{R}^n$ onto another which satisfies

$$(1) \quad \Phi(0) = 0$$
$$(2) \quad D_x\Phi(v) = \xi\bigl(\Phi(x)\bigr).$$

To begin with, note that any vector $x \in \mathbb{R}^n$ can be written uniquely in the form $x = \pi(x) + t(x)v$ where $t(x)$ is a real number, and π denotes orthogonal projection on the orthogonal complement of the line spanned by v. Observe that both π, t are linear mappings. We can suppose U chosen so that $\pi(U) \subseteq U$, $t(U) \subseteq I$. Define the smooth mapping $\Phi : U \to \mathbb{R}^n$ by

$$\Phi(x) = F\bigl(\pi(x), t(x)\bigr). \qquad\qquad *$$

Intuitively, to get to the point $\Phi(x)$ you start at the point $\pi(x)$ and slide up the flow-line through that point for time $t(x)$. The pictorial idea is as follows.

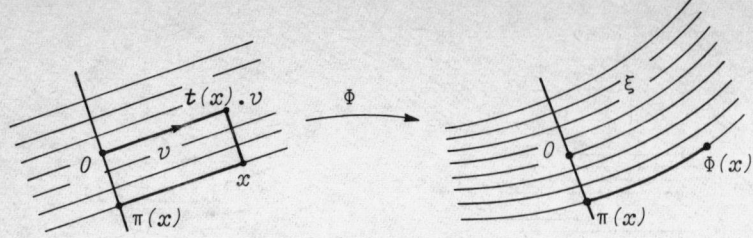

Certainly (1) is satisfied because one has $F(x, 0) = x$ for all $x \in U$. For (2) we proceed as follows. Taking differentials in *, and bearing in mind that π, t are linear, one obtains

$$D_x \Phi = \frac{\partial F}{\partial U}\bigl(\pi(x), t(x)\bigr) \circ \pi + \frac{\partial F}{\partial t}\bigl(\pi(x), t(x)\bigr) \circ t$$

with obvious meanings attached to $\frac{\partial F}{\partial U}$, $\frac{\partial F}{\partial t}$. The fact that $F(x, 0) = x$ for all $x \in U$ implies that $\frac{\partial F}{\partial U}(x, 0)$ is the identity map on $\mathbb{R}^n$. Note also that $\frac{\partial F}{\partial t}(0, 0)$ represents multiplication of the fixed vector v by a scalar. Thus $D_0 \Phi(x) = \pi(x) + t(x)v = x$ so $D_0 \Phi$ is the identity mapping on $\mathbb{R}^n$. It follows from the Inverse Function Theorem that Φ will map its domain U diffeomorphically onto its image, provided U is small enough. And that completes the proof. □

There is one last point which we should clear up, namely that much of what we have said can be extended to an apparently more general situation. In practice one often comes across "time-dependent" vector fields, i.e. one-parameter families of smooth vector fields. Formally, a <u>time-dependent</u> vector field on a smooth manifold N is a smooth mapping $\xi : N \times J \to TN$, with J an open interval, such that for each $t \in J$ the mapping $\xi_t : N \to TN$ defined by $\xi_t(x) = \xi(x, t)$ is a smooth vector field on N. Thus one thinks of ξ as a 1-parameter family $(\xi_t)_{t \in J}$ of smooth vector fields on N.

To maintain the contrast one can then refer to an ordinary vector field as <u>time-independent</u>.

For a time-dependent vector field $\xi : N \times J \to TN$ one can mimic the definitions already introduced for time-independent vector fields. Thus a <u>flow-line</u> for ξ starting at $x_0 \in N$ is a smooth mapping $f : I \to N$ with I an open interval containing 0, and $f(0) = x_0$, for which

$$f'(t) = \xi\bigl(f(t), t\bigr).$$

And a <u>smooth local flow</u> for ξ at x_0 is a smooth mapping $F : U \times I \to N$ (where U is a relatively open neighbourhood of x_0 in N, and I is an open interval containing 0) with the property that for any point $x \in U$ the map $f_x : I \to N$ defined by the rule $f_x(t) = F(x, t)$ is a flow-line for ξ starting at x. The analogue of (4.2) is

(4.3)   <u>Let ξ be a time-dependent vector field on a smooth manifold N, and let $x_0 \in N$: then there exists a smooth local flow for ξ at x_0</u>.

§5.   Germs of Smooth Mappings

The subject matter of this book is concerned principally with the behaviour of a smooth mapping $N \to P$ close to a point in its domain, where N, P are smooth manifolds. We can make this precise as follows. Suppose given a point $x \in N$, and consider the set of all smooth mappings $U \to P$ whose domain U is a neighbourhood of x in N. On this set we introduce an equivalence relation $\sim$. Given two such mappings $f_1 : U_1 \to P$, $f_2 : U_2 \to P$ we write $f_1 \sim f_2$ when there exists a neighbourhood U of x in N

depending on f_1 and f_2 for which the restrictions $f_1|U$, $f_2|U$ coincide. The equivalence classes under this relation are called <u>smooth germs</u> of mappings $N \to P$ at x, and elements of the equivalence class are called <u>representatives</u> of the germ. Notice that if f_1, f_2 are representatives of the same germ then $f_1(x) = f_2(x)$, so all representatives of the germ take the same value y, say, at x: in view of this fact it is usual to adopt the notation $f : (N, x) \to (P, y)$ for the germ, and to call x, y respectively the <u>source</u>, <u>target</u> of the germ. In the particular case when $N = P$ we use the notation $1_N : (N, x) \to (N, x)$ for the germ at x of the identity mapping $N \to N$.

One can handle germs in much the same way as one handles the mappings from which they are derived. For instance given germs $f : (N, x) \to (P, y)$ and $g : (P, y) \to (R, z)$ we can "compose" them to obtain a germ $g \circ f : (N, x) \to (R, z)$: one just chooses representatives $f : U \to P$, $g : V \to R$ with $f(U) \subseteq V$, which is evidently possible, and then the germ of their composite $g \circ f : U \to R$ at x can be defined to be the composite germ. It is a trivial matter to check that this definition does not depend on the particular choices of representatives. Likewise, one can pursue the analogy with mappings to introduce "inverses". A germ $f : (N, x) \to (P, y)$ is <u>invertible</u> when there exists a germ $g : (P, y) \to (N, x)$ for which one has $f \circ g = 1_P$, $g \circ f = 1_N$: and in that case g is called the <u>inverse</u> of f. Further, to a germ $f : (N, x) \to (P, y)$ we associate a differential, denoted as one would expect by $T_x f : T_x N \to T_y P$, and defined to be the differential at x of any representative: once again the definition does not depend on the choice of representative. And we shall leave the reader to write out in full for himself the obvious version of the Chain Rule for germs. It is perhaps worth pointing out that the Inverse Mapping Theorem admits a neat statement in the language of germs, namely that a germ is

invertible if and only if its differential is invertible. The <u>rank</u> of a germ
$f : (N, x) \to (P, y)$ is defined to be that of its differential: when the
rank equals $\dim N$ the germ is <u>immersive</u>, and when it equals $\dim P$ it is
<u>submersive</u>. Thus a germ will be invertible if and only if it is both immersive and submersive. A germ which is neither immersive nor submersive is
called <u>singular</u>.

<u>Example 1</u> Consider the plane curve $f : \mathbb{R} \to \mathbb{R}^2$ defined by $t \to (t^2, t^3)$.
Its image is the cuspidal cubic $x^3 = y^2$ pictured below. Clearly, the germ
at any point $t \neq 0$ is immersive, whilst that at $t = 0$ is singular. Thus
the only value of t for which we get a singular germ corresponds to the one
rather exceptional point on the curve, namely the cusp point.

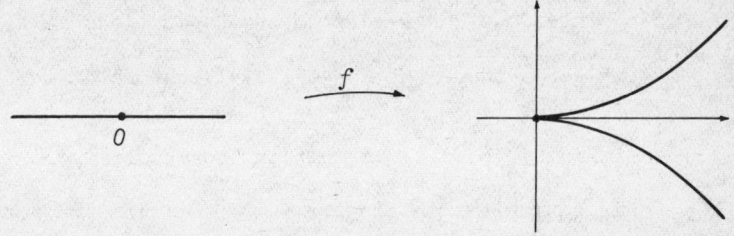

There are various equivalence relations under which it is sensible to study
smooth germs. A good starting point from which to develop is the following.
By analogy with the definition of equivalence given for mappings in §2 we call
two germs f_1, f_2 <u>equivalent</u> when there exist invertible germs h, k for
which the following diagram commutes

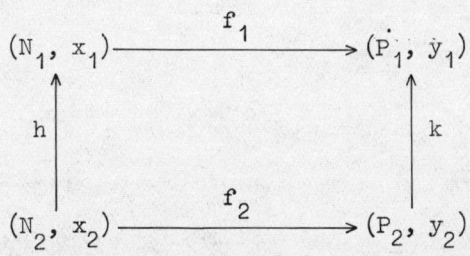

<u>Example 2</u> Consider the germs $(\mathbb{R}^2, 0) \to (\mathbb{R}^2, 0)$ given by the formulae

$$f_1(x, y) = (x^2 + y^2, xy) \qquad f_2(x, y) = (x^2, y^2).$$

These germs are equivalent because in the above diagram we can take

$$h(x, y) = (x + y, x - y) \qquad k(x, y) = (2x + 2y, x - y)$$

which are invertible because the formulae represent vector space isomorphisms, hence diffeomorphisms.

It is maybe worthwhile spelling out one simple consequence of the above definition, namely that any germ $f : (N, x) \to (P, y)$ is equivalent to some germ $(\mathbb{R}^n, 0) \to (\mathbb{R}^p, 0)$. It is for this reason that much of this book is restricted to studying germs of this apparently special type. Our broad objective is to introduce the reader to the basic ideas relevant to the problem of classifying germs under the above equivalence relation.

<u>Example 3</u> Proposition (1.3) tells us that any submersive germ $(\mathbb{R}^n, 0) \to (\mathbb{R}^p, 0)$ is equivalent to the germ at 0 of the projection $(x_1, \ldots, x_n) \to (x_1, \ldots, x_p)$. And Proposition (1.4) tells us that any immersive germ $(\mathbb{R}^n, 0) \to (\mathbb{R}^p, 0)$ is equivalent to the germ at 0 of the natural inclusion $(x_1, \ldots, x_n) \to (x_1, \ldots, x_n, 0, \ldots, 0)$.

This example takes care of the non-singular germs. What of the singular ones? Here one is up against serious mathematical difficulties, and it will take us to the final chapter of this book even to outline how one sets about resolving these difficulties.

Finally, we wish to introduce "jets" of smooth mappings, to be thought of as finite approximations to germs of smooth mappings, in the following rough sense: given a germ we can write down a corresponding Taylor series (of some representative with respect to appropriate local co-ordinates at source and target) and the jets of the germ correspond to the initial finite segments of the Taylor series. Let us make this more precise. By the **jet-space** $J^k(n, p)$ we mean the real vector space of all mappings $f : \mathbb{R}^n \to \mathbb{R}^p$ each of whose components is a polynomial of degree $\leq k$ in the standard co-ordinates $x_1, \ldots, x_n$ in $\mathbb{R}^n$ with zero constant term: the elements of $J^k(n, p)$ will be called **k-jets**. Suppose now that $f : \mathbb{R}^n \to \mathbb{R}^p$ is a smooth mapping, and that $a \in \mathbb{R}^n$. If in the Taylor series of $f(x) - f(a)$ at the origin (expressed in terms of the standard co-ordinates on $\mathbb{R}^n$, $\mathbb{R}^p$) we delete all terms of degree $> k$ the result can be thought of as a k-jet, which one writes $j^k f(a)$ and calls the **k-jet of (the germ of)** f **at** a. In this way we arrive at a smooth mapping $j^k f : \mathbb{R}^n \to J^k(n, p)$ given by $a \to j^k f(a)$ called the **k-jet extension of** f: this mapping will play a crucial role in the final sections of the next chapter.

II Transversality

§1. The Notion of Transversality

The notion of objects intersecting transversally (or in general position) has become quite fundamental to singularity theory. The simplest situation to look at is two subspaces of a vector space V: we say that they <u>intersect transversally</u> when their vector sum is V.

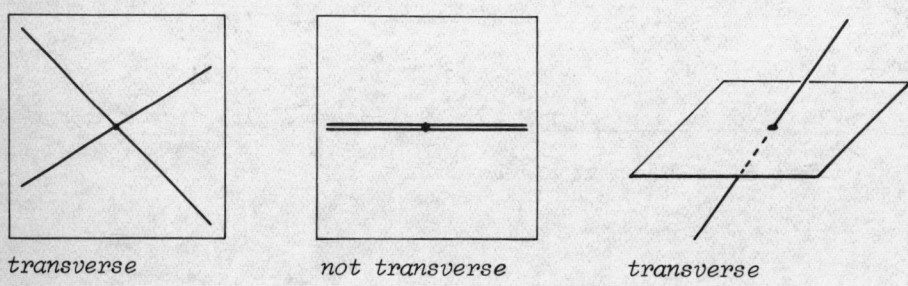

transverse   *not transverse*   *transverse*

The notion is easily extended to smooth submanifolds of a smooth manifold. We say that two smooth submanifolds N_1, N_2 of a smooth manifold N <u>intersect transversally at</u> $x \in N_1 \cap N_2$ when the tangent spaces $T_x N_1$, $T_x N_2$ intersect transversally in $T_x N$: and N_1, N_2 <u>intersect transversally in N</u> when they do so at every point in $N_1 \cap N_2$. Probably the best way to understand the idea is to look at a series of pictures.

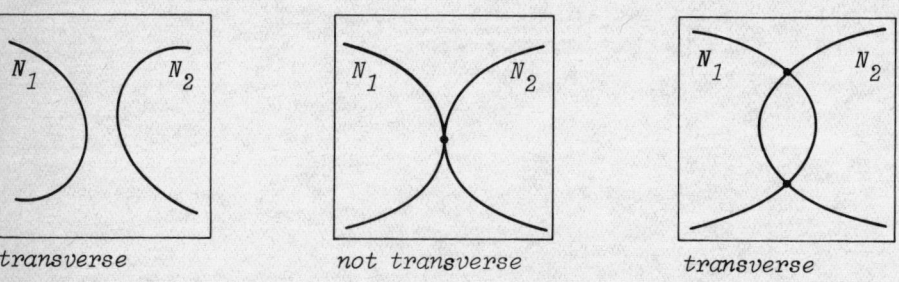

transverse   *not transverse*   *transverse*

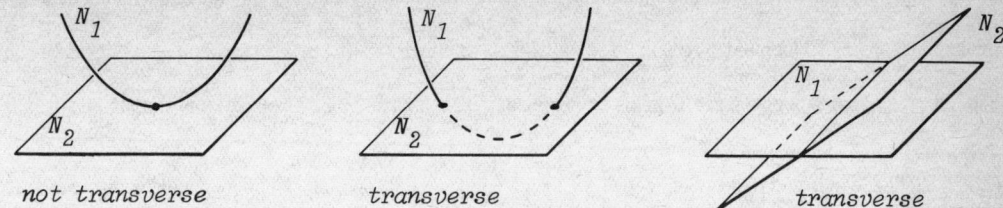

not transverse   transverse   transverse

The idea which we really want to exploit is this. Let $f : N \to P$ be a smooth mapping, and let Q be a smooth submanifold of P. Recall that both graph f and $N \times Q$ are smooth submanifolds of $N \times P$. We say f is transverse to Q and (write $f \pitchfork Q$) when graph f, $N \times Q$ intersect transversally in $N \times P$. One pictures it thus.

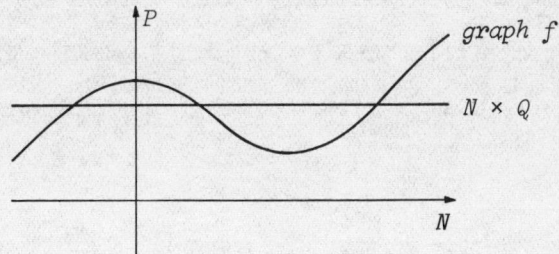

(1.1)   <u>Let $f : N^n \to P^p$ be a smooth mapping, and let Q be a smooth submanifold of P. An equivalent condition for $f \pitchfork Q$ is that for all $x \in N$ with $y = f(x) \in Q$ we have</u>

$$T_x f(T_x N) + T_y Q = T_y P . \quad\quad *$$

Proof   The condition for graph f, $N \times Q$ to intersect transversally in $N \times P$ is that for all points $z = (x, y)$ in the intersection, i.e. all points $(x, f(x))$ with $f(x) \in Q$, we have

$$T_z(\text{graph } f) + T_z(N \times Q) = T_z(N \times P).$$

In view of (I.3.2) and (I.3.3) this may be re-written as

$$\text{graph } T_x f + T_x N \times T_y Q = T_x N \times T_y P$$

which is clearly equivalent to *.   □

Some authors use the relation * as a definition of $f \pitchfork Q$, and certainly in practice * is probably easier to work with. However we prefer the definition given above on the ground that it has more immediate geometric content. There are some special cases well worthy of separate mention.

<u>Case 1</u>   Take the case when f is a <u>submersion</u> i.e. its germ at any point is submersive. In that case $T_x f\,(T_x N)$ is a p-dimensional subspace of the p-dimensional vector space $T_y P$, hence equal to it. We conclude that a submersion $f : N \to P$ must be transverse to every submanifold $Q \subseteq P$.

<u>Case 2</u>   A further particular situation is prompted by the observation that if * holds (for some x) then certainly codim $Q \leq$ dim N. Consequently if codim $Q >$ dim N then transversality of $f : N \to P$ to Q is equivalent to the image $f(N)$ being disjoint from Q, or (expressed more vividly) to f avoiding Q.

<u>Case 3</u>   One last special case is when Q comprises a single point (see the diagram above). A point in P to which f is transverse is called a <u>regular value</u> of f. A point $x \in N$ for which $f(x)$ fails to be a regular value is a <u>critical point</u> of f, and $f(x)$ is a <u>critical value</u>. Clearly, the condition for this is that $T_x f$ should fail to be surjective, i.e. have rank $< n$. For instance, when $N = P = \mathbb{R}$ the critical points are precisely the points where the derivative vanishes; and the critical values are precisely the real numbers c for which the line $y = c$ fails to intersect the graph of f transversally.

It is maybe worthwhile spelling out the fact that the notion of transversality is invariant under equivalence of smooth mappings, in the following precise sense. Suppose the smooth mappings f_1, f_2 are equivalent, i.e. one has a commuting diagram of smooth mappings

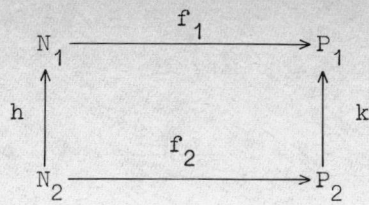

with h, k diffeomorphisms; and suppose further that Q_1, Q_2 are smooth submanifolds of P_1, P_2 respectively corresponding under k. The statement of the invariance of transversality is that f_1 is transverse to Q_1 if and only if f_2 is transverse to Q_2. The proof is a straightforward deduction from the definitions, and is left as an exercise for the reader. Bear this point in mind when reading the proof of the next proposition, which provides us with a painless procedure for extracting further examples of smooth manifolds.

(1.2)   Let $f : N^n \to P^p$ be a smooth mapping, and let Q^q be a smooth submanifold of P with $f \pitchfork Q$: then $M = f^{-1}Q$ is a smooth submanifold of N having the same codimension as Q, or is empty. Further, for any point x in N with $y = f(x)$ in Q one has $T_x M = T_x f^{-1}(T_y Q)$.

Step 1   For the first proposition it suffices to show that every point $x \in M$ has a neighbourhood whose intersection with M is a smooth manifold of codimension $r = p - q$. Put $y = f(x)$. By taking charts at x, y we can suppose that N, P are open sets in $\mathbb{R}^n$, $\mathbb{R}^p$ and that $x = 0$, $y = 0$: indeed (I.2.2) allows us to assume further that Q is the intersection of P with $\mathbb{R}^q \times 0$. Let π denote the projection of $\mathbb{R}^p$ on $0 \times \mathbb{R}^r$. The transversality of f to Q tells us that $\pi \circ f$ has rank r at 0, and then (I.1.3) tells us that there is a diffeomorphism k of a neighbourhood of 0 in $\mathbb{R}^n$ onto another with $\pi \circ f \circ h$ the projection $(x_1, \ldots, x_n) \to (x_1, \ldots, x_r)$. The inverse image of 0 under this mapping is an open subset of $0 \times \mathbb{R}^{n-r}$, corresponding under h to M: that shows that M is a smooth manifold of codimension r.

<u>Step 2</u> The tangent space T_xM is computed as follows. The commuting diagram of smooth mappings on the left gives rise to the commuting diagram of differentials on the right.

from which it is clear that $T_xM \subseteq T_xf^{-1}(T_yQ)$. To show that these vector spaces are equal it will suffice to show that their dimensions are equal. For this one considers the linear mapping $T_xf^{-1}(T_yQ) \to T_yQ$ given by restricting T_xf: note that it has the same kernel as T_xf, and image $T_yQ \cap T_xf(T_xN)$. The desired equality now follows on using the fact that the dimension of the domain is the sum of the dimensions of kernel and image, together with the definition of transversality. □

Note the special case of this result when Q is a single point y. In that case the hypothesis is that y is a regular point of f; the conclusion is that $f^{-1}(y)$ is a smooth manifold of dimension $n - p$, or empty, and that T_xM is the kernel of T_xf.

<u>Example 1</u> Let $A = (a_{ij})$ be an invertible symmetric $n \times n$ matrix. Consider the smooth mapping $f : \mathbb{R}^n \to \mathbb{R}$ which is given by $f(x) = Ax.x$, where . denotes the standard scalar product on $\mathbb{R}^n$. Here the differential at the point x is the linear mapping $v \to 2(Ax.v)$, so 1 is a regular value of f, and $f^{-1}(1)$, i.e. the central quadric $\sum a_{ij}x_ix_j = 1$, is a smooth manifold of dimension $(n - 1)$, or empty. And the tangent space

at x to this quadric is the kernel of the differential, i.e. the hyperplane perpendicular to the vector Ax. Taking A to be the identity matrix we see that S^{n-1} is a smooth submanifold of $\mathbb{R}^n$ of dimension $(n-1)$, a fact we already knew from Chapter I: moreover the tangent space to this sphere at a point x is precisely the hyperplane perpendicular to x, as we would expect. Taking $n = 3$ we see that ellipsoids, hyperbolic cylinders, elliptic cylinders, and hyperboloids are all smooth manifolds of dimension 2.

Example 2   Consider a smooth mapping $f : \mathbb{R}^2 \to \mathbb{R}$ whose differential has rank 1 at every point in $f^{-1}(0)$: (1.2) tells us that $f^{-1}(0)$ is then a smooth submanifold of $\mathbb{R}^2$ of dimension 1, i.e. a smooth curve in the plane. Consider now $F : \mathbb{R}^3 - \{0\} \to \mathbb{R}$ given by $F(x, y, z) = f(\sqrt{x^2 + y^2}, z)$. This too has the property that the differential has rank 1 at every point in $F^{-1}(0)$, so likewise $F^{-1}(0)$ is a smooth submanifold of $\mathbb{R}^3$ of dimension 2: it is just the surface of revolution obtained by rotating the curve $f(y, z) = 0$, $y > 0$ about the z-axis in $\mathbb{R}^3$. By way of specific illustration let a, b be real numbers with $a > b > 0$, and let f be given by

$$f(y, z) = (y - a)^2 + z^2 - b^2.$$

Clearly the differential of f has rank 1 at every point in $f^{-1}(0)$, which is the circle radius b centred at $(a, 0)$. The surface of revolution $F^{-1}(0)$ is a torus, and a minor computation will verify that it is given by the equation

$$(x^2 + y^2 + z^2 + a^2 - b^2)^2 = 4a^2(x^2 + y^2).$$

Example 3   Suppose A^a, B^b are smooth submanifolds of a smooth manifold C^c intersecting transversally in C: the intersection $A \cap B$ will be empty, or a smooth submanifold of C of dimension $a + b - c$. This follows on observing that the inclusions $A \to C$, $B \to C$ are transverse to B, A respectively.

A <u>singular point</u> (or <u>singularity</u>) of a smooth mapping $f : N \to P$ is a point $x \in N$ where the germ is singular. (Of course a singular point is a critical point, though the converse does not necessarily hold.) The <u>singular set</u> Σf of f is the set of all its singular points: the image of Σf is sometimes called the <u>bifurcation set</u>.

Example 4 The Whitney cusp mapping of the plane is the smooth mapping $f : \mathbb{R}^2 \to \mathbb{R}^2$ given by $(x, y) \to (u, v)$ where $u = x$, $v = y^3 - xy$. The singular set is the set of points where the Jacobian matrix has rank < 2, i.e. the parabola $x = 3y^2$. And the bifurcation set is the image of this parabola under f, i.e. the cuspidal cubic having the equation $4u^3 - 27v^2 = 0$.

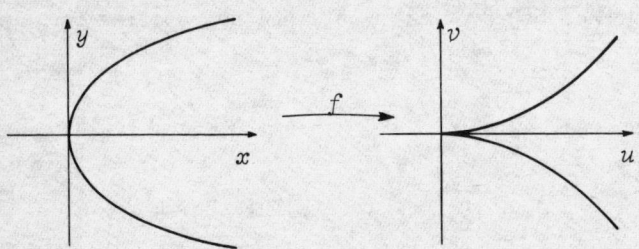

The Whitney cusp mapping is best understood on a geometric level as the composite of the mapping $g : \mathbb{R}^2 \to \mathbb{R}^3$ given by $(x, y) \to (u, v, w)$ with $u = x$, $v = y^3 - xy$, $w = y$ and the projection $\pi : \mathbb{R}^3 \to \mathbb{R}^2$ given by $(u, v, w) \to (u, v)$: see the diagram below. The image of g is the folded surface S defined by $v - w^3 + uw = 0$, a smooth submanifold of $\mathbb{R}^3$ of dimension 2 by (1.2).

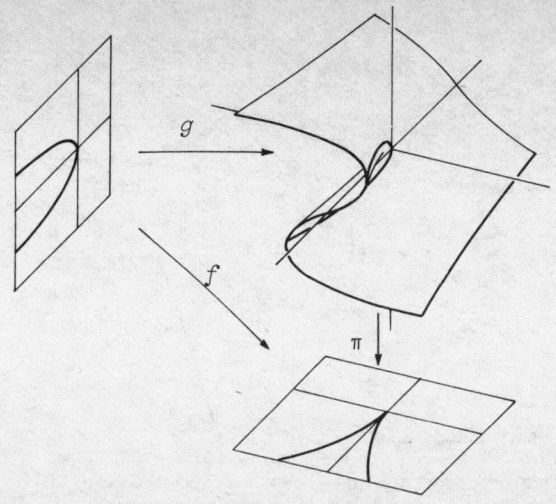

It is maybe worth saying a little more about the projection of a surface onto a plane, as it helps to strengthen one's intuition. Consider a smooth surface $S \subseteq \mathbb{R}^3$, i.e. a 2-dimensional smooth submanifold of $\mathbb{R}^3$, and a plane $P \subseteq \mathbb{R}^3$ through the origin: we take π to denote orthogonal projection onto P, and ask for the singular set of the restriction $\pi|S$. Think of P as "horizontal", and the line perpendicular to it as "vertical". For each point $x \in S$ the commuting diagram of smooth mappings on the left gives rise to a commuting diagram of differentials on the right.

The condition for x to be a singular point of $\pi|S$ is therefore that π should map $T_x S$ onto a proper subspace of P, i.e. that the tangent plane $T_x S$ should be "vertical"; thus, in general, the singular set of $\pi|S$ will be a curve in S projecting to a curve in P, the bifurcation set.

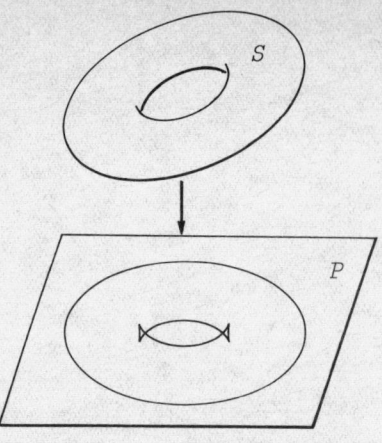

Example 5   Let us return to the model situation presented by the last example. The folded surface was given by the equation $F = v - w^3 + uw = 0$, and was projected onto the plane $w = 0$. The tangent plane to S is the kernel of the differential of F, i.e. the plane perpendicular to the vector $\left(\frac{\partial F}{\partial u}, \frac{\partial F}{\partial v}, \frac{\partial F}{\partial w}\right)$, i.e. to $(w, 1, u - 3w^2)$: and the tangent plane will be "vertical" when this vector lies in the plane $w = 0$, i.e. when $u = 3w^2$, $v = -2w$. Thus the singular set of $\pi | S$ is the "fold curve" parametrized by the equations $u = 3t^2$, $v = -2t^3$, $w = t$; the geometrically inclined reader will recognise this as a twisted cubic, projecting onto the cuspidal cubic. Note that one point on the "fold curve", namely that projecting to the cusp, differs qualitatively from the rest; it is the one point where two folds on the surface meet, whereas all the others are just points where the surface folds over.   In some sense therefore we have two types of singular point.

Example 6   Consider the projection π of the torus S of Example 2 onto the plane $y = z$. Recall that S was given by an equation

$$G = (x^2 + y^2 + z^2 + a^2 - b^2)^2 - 4a^2(x^2 + y^2) = 0. \qquad (1)$$

The tangent plane to S is the kernel of the differential of G, i.e. the plane perpendicular to the vector $\left(\frac{\partial G}{\partial x}, \frac{\partial G}{\partial y}, \frac{\partial G}{\partial z}\right)$: and the condition for a point to be a singular point of $\pi|S$ is that this vector should lie in the plane $y = z$, i.e. that $\frac{\partial G}{\partial y} = \frac{\partial G}{\partial z}$, which computation verifies to be the condition that

$$(z - y)(x^2 + y^2 + z^2 + a^2 - b^2) + 2a^2 y = 0. \qquad (2)$$

The singular set of $\pi|S$ is given therefore by (1), (2), i.e. it is the intersection of two surfaces of degrees 4, 3 respectively, hence a space curve of degree 12. And its projection onto the plane $y = z$ is the bifurcation set; it is what we "see" if we imagine the torus made of glass and viewed from a distant point on the line perpendicular to the plane $y = z$.

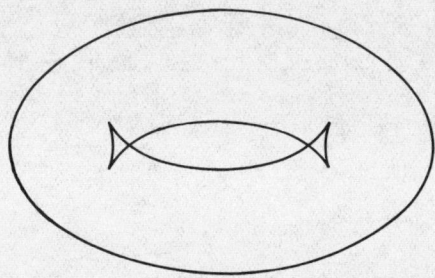

So much for transversality per se. The remainder of this chapter is devoted to the development of a very basic intuition associated with transversality. Suppose we have a smooth mapping $f : N \to P$ which <u>fails</u> to be transverse to a submanifold $Q \subseteq P$. Common sense tells one that f can be forced to be transverse to Q by arbitrarily small perturbations. The formal development of this intuition leads to a host of results, reaching high degrees of complexity and subtlety, called transversality theorems.

In recent years transversality theorems have assumed a role of increasing importance in differential topology, indeed Singularity Theory could hardly exist without them. Essentially they all say the same thing, namely that by an arbitrarily small perturbation a given smooth mapping (or some closely related smooth mapping) can be made transverse to a smooth manifold. The object of the next section is to establish a very basic lemma which provides the key to most transversality theorems.

§2. The Basic Transversality Lemma

Let $F : N \times S \to P$ be a smooth mapping. This we think of as a smooth family of smooth maps $f_s : N \to P$ where $f_s(x) = F(x, s)$ parametrized by the elements $s \in S$. Suppose $F \pitchfork Q$ with Q a smooth submanifold of P. We ask whether $f_s \pitchfork Q$ for all parameters s? That the answer can well be in the negative is shown by

<u>Example 1</u>   Take $N = \mathbb{R}^2$, $S = \mathbb{R}$, $P = \mathbb{R}^3$ and consider the smooth family $F : N \times S \to P$ defined by the formula $((x, y), s) \to (x, y, s)$. Thus f_s maps $\mathbb{R}^2$ onto the horizontal plane $z = s$ in $\mathbb{R}^3$. And take Q to be the 2-sphere S^2. F is a diffeomorphism, so $F \pitchfork Q$. On the other hand $f_s \pitchfork Q$ only provided we avoid the two exceptional parameters $s = \pm 1$ when the plane $z = s$ is tangent to Q.

However, it is clear in this example that any value of the parameter s can be approximated as closely as we please by "good" values, i.e. those for which $f_s \pitchfork Q$. That this holds generally is the content of the Basic Transversality Lemma: for technical reasons we prefer to replace the single smooth manifold Q by several smooth manifolds $Q_1, \ldots, Q_t$: the reason will become clear in the final section of this chapter. The crucial tool here is Sard's Theorem, one of the truly fundamental results in differential topology, which we state in the following form.

(2.1)   Let $f_i : N_i \to P$ be a countable family of smooth mappings. The set of common regular values of the f_i is dense in P.

The proof of Sard's Theorem is quite lengthy, so we have isolated it as Appendix A to this book in order not to break the flow of the text. The Basic Transversality Lemma is

(2.2)   Let $F : N \times S \to P$ be a smooth family of smooth mappings transverse to smooth submanifolds $Q_1, \ldots, Q_t$ of P: then there is a dense set of parameters s for which f_s is transverse to all of $Q_1, \ldots, Q_t$.

Proof   By (1.2) $M_i = F^{-1}(Q_i)$ is a smooth submanifold of $N \times S$. Consider the restriction $\pi|M_i$ of the projection $\pi : N \times S \to S$. We shall show that

$$\pi|M_i \pitchfork \{s\} \text{ if and only if } f_s \pitchfork Q_i \quad\quad\quad *$$

which will prove the result, by Sard's Theorem. Now to the proof of $*$. We can, and shall, drop the index i since it plays no further role. The reader is urged to keep the following picture in mind.

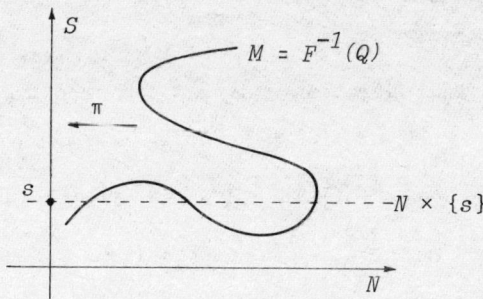

To start with, note that the condition for F to be transverse to Q is that for all $z = (x, s)$ in $N \times S$ with $w = F(z)$ in Q we have

$$T_z F(T_x N \times T_s S) + T_w Q = T_w P. \quad\quad\quad (1)$$

49

Consider now the condition for f_s to be transverse to Q: this is the condition that for the same x as in (1)

$$T_z F(T_x N \times 0) + T_w Q = T_w P. \quad\quad\quad (2)$$

Furthermore, the condition for $N \times \{s\}$ to be transverse to M is that for the same x as in (1)

$$T_x N \times 0 + T_z M = T_x N \times T_s S \quad\quad\quad (3)$$

where $T_z M = T_z F^{-1}(T_w Q)$ by (1.2). If we assume (3) and apply $T_z F$ to both sides then (1) tells us that (2) holds. And conversely it is a minor exercise in linear algebra to see directly that (2) implies (3). Thus the condition for f_s to be transverse to Q is precisely the condition for $N \times \{s\}$ to be transverse to M. We claim that this in turn is equivalent to the condition that $\pi|M$ is transverse to $\{s\}$: indeed this last condition says that for the same x as in (1) we have

$$T_z \pi(T_z M) = T_s S \quad\quad\quad (4)$$

which is clearly equivalent to (3). □

By way of explicit illustration consider two smooth manifolds M, N in some Euclidean space S. Even if they do not intersect transversally, geometric intuition tells us that it should be possible to force them to do so by an arbitrarily small translation of one manifold M in some direction. We can make this feeling precise as follows. For $s \in S$ let M_s denote the image of M under the translation of S defined by $x \mapsto x + s$. We claim that the s for which M_s, N intersect transversally will be dense in S, justifying one's intuition. To this end define a smooth family of smooth mappings $F: M \times S \to S$ by $(x, s) \mapsto x + s$. It is clear that F is a submersion, so F is transverse to N. By the Basic Transversality Lemma f_s is transverse to N for a dense set of parameters s. Our claim is established by observing that M_s, N intersect transversally if and only if f_s is transverse to N — which fact we leave as an exercise for the reader.

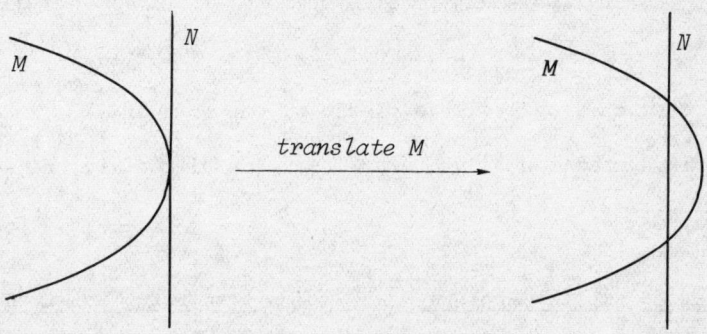

§3. An Elementary Transversality Theorem

Our object in this section will be to show that given a smooth mapping $f : \mathbb{R}^n \to \mathbb{R}^p$, and a smooth manifold $Q \subseteq \mathbb{R}^p$, we can find a smooth mapping $g : \mathbb{R}^n \to \mathbb{R}^p$ which is transverse to Q, and as "close" as we please to f. Of course the first problem is to say precisely what we mean by smooth mappings being "close": roughly speaking we shall take this to mean that their values are "close", and that for each integer $k \geq 1$ their derivatives of order k are "close".

We make this precise as follows. Let $C^\infty(\mathbb{R}^n, \mathbb{R}^p)$ denote the set of all smooth mappings $\mathbb{R}^n \to \mathbb{R}^p$, and let $f : \mathbb{R}^n \to \mathbb{R}^p$ be a given smooth mapping. Given a (small) positive real number ϵ, a (large) positive real number R, and an integer $k \geq 0$ we associate to f a <u>fundamental neighbourhood</u> in $C^\infty(\mathbb{R}^n, \mathbb{R}^p)$ comprising all those smooth mappings $g : \mathbb{R}^n \to \mathbb{R}^p$ for which for all $x \in \mathbb{R}^n$ with $|x| \leq R$ one has

$$\|j^k f(x) - j^k g(x)\| \leq \epsilon$$

with $\| \ \|$ a fixed norm on the jet-space $J^k(\mathbb{R}^n, \mathbb{R}^p)$. And we call a subset $X \subseteq C^\infty(\mathbb{R}^n, \mathbb{R}^p)$ <u>dense</u> therein when given any smooth mapping $f : \mathbb{R}^n \to \mathbb{R}^p$ and any fundamental neighbourhood V of f one can find a smooth mapping $g : \mathbb{R}^n \to \mathbb{R}^p$ in X with $g \in V$: intuitively, any mapping can be approximated as closely as we please by mappings in X.

We are now in a position to state and prove an elementary transversality theorem.

(3.1) <u>The set of smooth mappings $\mathbb{R}^n \to \mathbb{R}^p$ transverse to given smooth submanifolds $Q_1, \ldots, Q_t$ of $\mathbb{R}^p$ is dense in</u> $C^\infty(\mathbb{R}^n, \mathbb{R}^p)$.

<u>Proof</u> Let $f : \mathbb{R}^n \to \mathbb{R}^p$ be smooth. We have to show that we can approximate f as closely as we please by mappings transverse to $Q_1, \ldots, Q_t$. The idea is to construct a smooth family $F : \mathbb{R}^n \times S \to \mathbb{R}^p$ which contains f, and with F transverse to $Q_1, \ldots, Q_t$: then one applies the Basic Transversality Lemma. We shall make F transverse to $Q_1, \ldots, Q_t$ by ensuring that it is a submersion. To motivate the construction recall that the transversality of $f : \mathbb{R}^n \to \mathbb{R}^p$ to Q is equivalent to the graph being transverse to $\mathbb{R}^n \times Q$ in the product $\mathbb{R}^n \times \mathbb{R}^p$. If the graph is not already transverse it seems reasonable that we might be able to force it to be so by

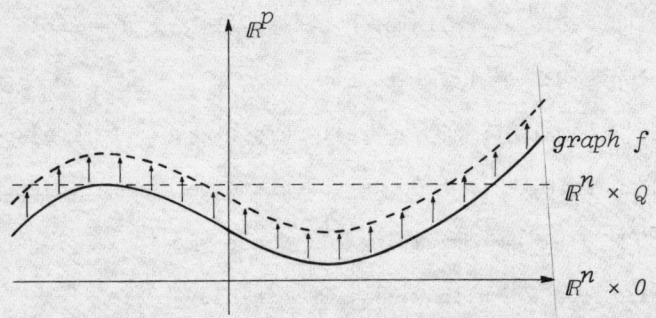

bodily translating it. The picture above illustrates the idea when Q is a point. Therefore we take $S = \mathbb{R}^p$ and define $F : \mathbb{R}^n \times \mathbb{R}^p \to \mathbb{R}^p$ by $(x, s) \mapsto f(x) + s$. Clearly, this is a submersion, so transverse to $Q_1, \ldots, Q_t$. By the Basic Transversality Lemma there is a dense set of s for which f_s is transverse to $Q_1, \ldots, Q_t$. All that remains to be shown is that if s is close enough to 0 then f_s is as close as we please to $f_0 = f$: more formally, we have to check that given a fundamental neighbourhood V of f we can find an $s \neq 0$ for which f_s lies in V. And that we can safely leave to the reader. □

It is only fair to point out that this example of a transversality theorem is not particularly useful. Its virtue lies rather in the fact that it is an easily understood prototype of transversality theorems of greater complexity and application. Our object here was simply to lay bare the underlying idea behind the proofs of such theorems.

§4. Thom's Transversality Theorem

Our next transversality theorem is rather more useful than that of the preceding section. Indeed it will suffice for all the applications we shall require in this book. Its statement does not possess quite the same immediate intuitive appeal as our previous result: however, in the next section we shall discuss a simple application which should clarify the situation.

(4.1) <u>Let $Q_1, \ldots, Q_t$ be smooth submanifolds of the jet space</u> $J^k(n, p)$. <u>The set of all smooth mappings</u> $f : \mathbb{R}^n \to \mathbb{R}^p$ <u>for which</u> $j^k f : \mathbb{R}^n \to J^k(n, p)$ <u>is transverse to</u> $Q_1, \ldots, Q_t$ <u>is dense in</u> $C^\infty(\mathbb{R}^n, \mathbb{R}^p)$.

<u>Note</u>   It is essential that the reader appreciate the difference between this transversality theorem and that of the preceding section. In the previous result we managed to make f transverse to a submanifold by using a <u>constant</u> deformation. But in the present situation it is $j^k f$, not f, which we wish to make transverse to a submanifold. On the other hand it is only f which we are allowed to deform. A constant deformation will not work here since it does not alter the derivatives of f. What we do instead is to use a <u>polynomial</u> deformation.

<u>Proof</u>   Let $S = J^k(n, p)$. Of course S is finite-dimensional, so can be identified with a Euclidean space. Consider the smooth family $F : \mathbb{R}^n \times S \to J^k(n, p)$ defined by $(x, s) \to j^k(f + s)(x)$. F is a submersion, because for fixed x the mapping represents an affine isomorphism of $J^k(n, p)$ with itself, hence transverse to all of $Q_1, \ldots, Q_t$. By the Basic Transversality Lemma there is a dense set of parameters s for which the mapping $f_s : \mathbb{R}^n \to J^k(n, p)$ defined by $x \to j^k(f + s)(x)$ is transverse to all of $Q_1, \ldots, Q_t$. And clearly we can make $f + s$ as close as we please to f by choosing s to be sufficiently small. □

§5. First Order Singularity Sets

This section is in the nature of an extended example. We shall use the Thom Transversality Theorem to show that for a dense set of smooth mappings $f : \mathbb{R}^n \to \mathbb{R}^p$ the singular set Σf can be partitioned into finitely many smooth manifolds on each of which f has constant rank.

Recall that a singular point of a smooth mapping $f : N^n \to P^p$ is a point $x \in N$ for which the rank of the differential $T_x f$ falls below its possible maximal value of min (n, p). A natural way of distinguishing one singular point from another is by the actual value taken by the rank of the differential. To this end we introduce the <u>first order Thom singularity sets</u>

$$\Sigma^i f = \{x \in N : T_x f \text{ has kernel rank } i\}.$$

(Concerning the terminology: later in this book we shall have something to say about higher order Thom singularity sets.) In this way we obtain a partition of N into finitely many sets on each of which g has constant rank. One might reasonably hope that these sets will be submanifolds of N, but that (as the following examples will illustrate) is not necessarily the case.

<u>Example 1</u> Take $f : \mathbb{R}^2 \to \mathbb{R}^2$ defined by the formula $(x, y) \to (x^2, y^2)$. This we shall refer to as the "folded handkerchief" mapping for the following reason: it is the composite of the two mappings

$$(x, y) \mapsto (x, y^2) : (x, y) \mapsto (x^2, y)$$

which "fold" the plane along the x-axis, y-axis respectively. One pictures it something like this.

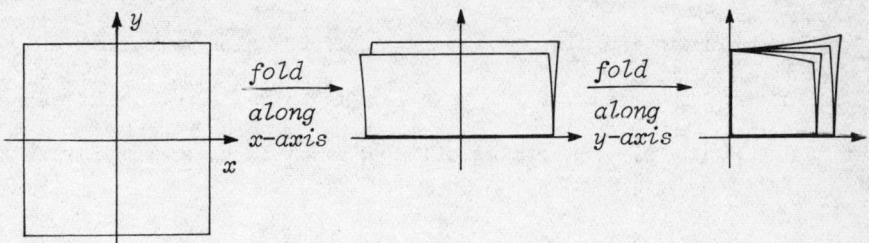

The reader can easily check that the origin is the only Σ^2-point, other points on the axes being Σ^1 points, and the rest being Σ^0 points. (One need hardly point out that this fits in with the fact that the origin is folded twice, other points on the axes just once, and the rest not at all.) Here the $\Sigma^i f$ are all submanifolds.

Example 2   Take the smooth mapping $f : \mathbb{R}^2 \to \mathbb{R}^2$ given by $(x, y) \mapsto (x^2 + y, y^2)$. A minor computation shows that the Σ^1 points are given by $xy = 0$, i.e. $\Sigma^1 f$ is the union of the two axes, so certainly not a submanifold. (There are no Σ^2 points.) However, it is well worth observing that in this example we can force $\Sigma^1 f$ to be a submanifold by slightly deforming f. Of course a <u>constant</u> deformation will not affect the Jacobian matrix, and is therefore of no use. But a small <u>linear</u> deformation will do the trick. Take for instance $f_s : \mathbb{R}^2 \to \mathbb{R}^2$ defined by $(x, y) \to (x^2 + y, y^2 + 4sx)$. Another minor computation shows that the Σ^1 points are now given by $xy = s$, a hyperbola, so certainly a submanifold. Thus as s moves from 0 to a small value so the singularity set $\Sigma^1 f_s$ changes from a pair of intersecting lines to a pair of disjoint curves.

We shall now prove generally that the first-order singularity set $\Sigma^i f$ of a smooth mapping $f : N \to P$ can be forced to be submanifolds, by the device of slightly deforming f. For the sake of technical simplicity we shall carry it through only in the case $N = \mathbb{R}^n$, $P = \mathbb{R}^p$. Suppose then that $f : \mathbb{R}^n \to \mathbb{R}^p$ is our smooth mapping. Notice that given a point $x \in \mathbb{R}^n$ the first order singularity set to which it belongs depends only on the 1-jet of f at x, i.e. $D_x f$. So let us write Σ^i for the subset of the jet space $J^1(n, p)$ comprising all 1-jets which have kernel rank equal to i.

(5.1) Σ^i <u>is a smooth submanifold of</u> $J^1(n, p)$ <u>of codimension</u> $i(p - n + i)$.

Proof   For this it will be convenient to identify a linear map with its matrix relative to the standard bases. We proceed in two steps.

Step 1   Let $E = \begin{pmatrix} A & B \\ C & D \end{pmatrix}$ be a $p \times n$ matrix with A an invertible $k \times k$ matrix. We claim that E has rank k if and only if $D = CA^{-1}B$. To see this, observe that for an arbitrary matrix X the matrix E has the same rank as

$$\begin{pmatrix} I_k & 0 \\ X & I_{p-k} \end{pmatrix} \begin{pmatrix} A & B \\ C & D \end{pmatrix} = \begin{pmatrix} A & B \\ XA + C & XB + D \end{pmatrix}$$

and the claim follows on choosing X such that $XA + C = 0$.

Step 2   Choose k with $i + k = n$, and let E_0 be a matrix in Σ^i. It will be no restriction to suppose $E_0 = \begin{pmatrix} A_0 & B_0 \\ C_0 & D_0 \end{pmatrix}$ with A_0 a $k \times k$ invertible matrix. And it will suffice to produce an open neighbourhood U of E_0 in $J^1(n, p)$ with $U \cap \Sigma^i$ a smooth submanifold of codimension $i(p - n + i)$. We take U to be an open neighbourhood of E_0 with the property that for any matrix $E = \begin{pmatrix} A & B \\ C & D \end{pmatrix}$ in it the $k \times k$ matrix A is invertible. Consider the smooth mapping $f : U \to J^1(p - k, n - k)$ given by $E \to D - CA^{-1}B$. Notice that f is a submersion: indeed if one fixes A, B, C the resulting function of D is an invertible affine mapping. It follows from (1.2) that $f^{-1}(0)$ is a smooth submanifold of codimension $(p - k)(n - k) = i(p - n + i)$: but by Step 1 the inverse image $f^{-1}(0) = U \cap \Sigma^i$, which ends the proof. □

(5.2)   <u>There is a dense set of smooth mappings</u> $f : \mathbb{R}^n \to \mathbb{R}^p$ <u>for which</u> $j^1 f$ <u>is transverse to all the sets</u> Σ^i, <u>and hence for which each</u> $\Sigma^i f$ <u>is a smooth manifold of codimension</u> $i(p - n + i)$.

Proof   The first statement is an immediate consequence of the Thom Transversality Theorem (taking $k = 1$). And the second statement follows from the fact that $\Sigma^i f$ is just the inverse image under $j^1 f$ of Σ^i, so is a smooth manifold of the same codimension by (1.2). □

Example 3   Consider a smooth mapping $f : \mathbb{R}^n \to \mathbb{R}^p$ for which $j^1 f$ is transverse to the Σ^i in the special case when $p \geq 2n$. In this case $\Sigma^i f$ will have codimension $i(p - n + i) \geq i(n + i) > n$, unless $i = 0$. Thus $\Sigma^i f$ is void for $i > 0$, which is the same thing as saying that f is an immersion i.e. its germ at any point is immersive. Our theory says then that when $p \geq 2n$ any smooth mapping $\mathbb{R}^n \to \mathbb{R}^p$ can be slightly deformed to become an immersion, and indeed a linear deformation will suffice (as is clear from the proof of the Thom Transversality Theorem). For an explicit instance take the cuspidal cubic curve $f : \mathbb{R} \to \mathbb{R}^2$ given by $f(t) = (t^2, t^3)$. Here, for a small positive number s the curve $f_s : \mathbb{R} \to \mathbb{R}^2$ given by $f_s(t) = (t^2, t^3 - ts)$ is an immersion, obtained from f by a small linear deformation.

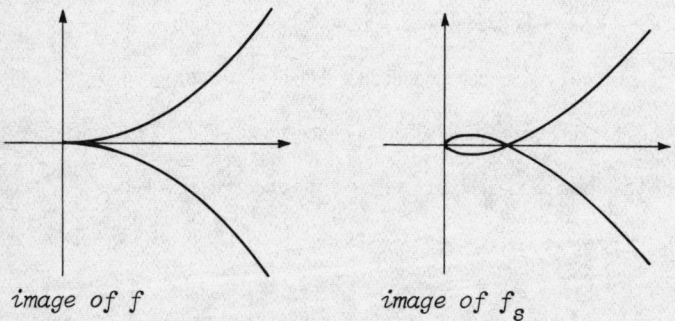

image of f　　　　*image of f_s*

Example 4   We shall consider, in some detail, the condition that $j^1 f$ be transverse to the Σ^i for a <u>function</u> $f : \mathbb{R}^n \to \mathbb{R}$. This means we are to have

$$\text{image of differential of } j^1 f \text{ at } a \; + \; \text{tangent space to } \Sigma^i \text{ at } j^1 f(a) \; = \; J^1(n, 1) \quad \text{———}*$$

for all points $a \in \mathbb{R}^n$. We shall show that this is exactly the condition that the so-called <u>Hessian</u> matrix

$$H = \left(\frac{\partial^2 f}{\partial x_i \partial x_j}(a)\right)_{n \times n}$$

is non-singular at every critical point a, where $x_1, \ldots, x_n$ are the standard co-ordinate functions in $\mathbb{R}^n$.

To start with, let us simplify matters by identifying $J^1(n, 1)$ with $\mathbb{R}^n$ by identifying a linear mapping $\mathbb{R}^n \to \mathbb{R}$ with its matrix relative to the canonical bases. In particular $D_a f$ is identified with $\left(\frac{\partial f}{\partial x_1}(a), \ldots, \frac{\partial f}{\partial x_n}(a)\right)$. Notice that the only Σ^i which arise are Σ^{n-1}, Σ^n of codimensions 0, n respectively. Thus Σ^{n-1} is an open set and $j^1 f$ is automatically transverse to it. $\Sigma^n f$ is defined by the n conditions $\frac{\partial f}{\partial x_1} = 0, \ldots, \frac{\partial f}{\partial x_n} = 0$ and is just the critical set of f. Thus we need only satisfy * when a is a critical point, and $i = n$. In that case the expressions in * are easily identified. The Jacobian matrix of $j^1 f$ at a is precisely H, so the image of the differential of $j^1 f$ at a is the subspace of $\mathbb{R}^n$ generated by its columns. Σ^n is just the origin in $\mathbb{R}^n$, so its tangent space can be discounted. Thus the burden of * is that the columns of H generate $\mathbb{R}^n$, i.e. that H is non-singular, establishing our claim.

A critical point of a smooth function $f : \mathbb{R}^n \to \mathbb{R}$ at which the Hessian matrix is non-singular is called <u>non-degenerate</u>. Our discussion can then be summed up by saying the condition for $j^1 f$ to be transverse to the Σ^i is that every critical point should be non-degenerate. By way of illustration, the reader may care to check that the function

$$f(x, y) = x^2 - 3xy^2$$

has a degenerate critical point at the origin. One pictures it like this.

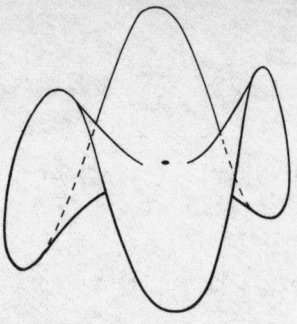

In this example j^1f cannot be transverse to the Σ^i. According to the theory of this section it should be possible to gain transversality to the Σ^i by a <u>linear</u> deformation of f, i.e. by taking $f_s = f + L_s$ where L_s is linear. Here is an explicit linear deformation which does just that — the reader is urged to check this for himself.

$$f_s(x, y) = x^3 - 3xy^2 - sx.$$

The graph of f_s is rather difficult to draw. However, some idea of what happens is indicated in the pictures below. The left-hand picture refers to the case $s = 0$, the right-hand one to the case $s > 0$. In both cases the shaded area is that where $f_s > 0$. And in both figures the curves $f_s = \pm\epsilon$ have been drawn for small positive values of ϵ.

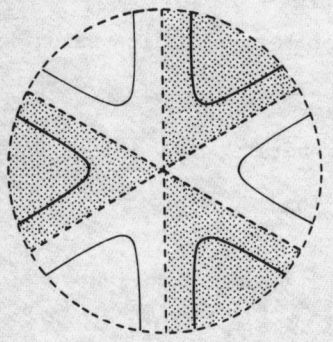
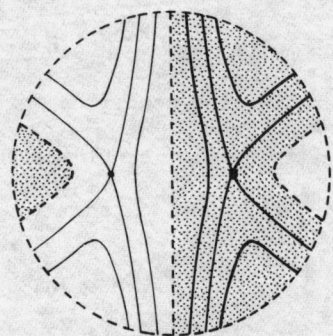

III Unfoldings: the finite dimensional model

The kind of mathematics which we shall discuss later in this volume lies rather deep, at least in the sense that formal proofs of the main results rely on hard theorems in analysis. However the underlying geometric ideas are rather straightforward. The main object of this chapter is to introduce these ideas in a relatively simple situation where the mathematics is easy enough not to hinder understanding: it will also provide us with an opportunity of establishing one or two little facts which we shall have occasion to use later.

§1. Groups Acting on Sets

By an <u>action</u> of a group G on a set M we mean a mapping $\Phi : G \times M \to M$, usually written $(g, x) \mapsto g.x$, for which for all $x \in M$, and $g, h \in G$

(i) $1.x = x$

(ii) $(gh).x = g.(h.x)$

where 1 denotes the identity of G.

Given such an action we can define an equivalence relation $\sim$ on M by agreeing that $x \sim y$ when there exists an element $g \in G$ for which $y = g.x$. The equivalence classes are called the <u>orbits</u> under the action. Given $x \in M$ the <u>orbit through</u> x is by definition the equivalence class which contains x, i.e. the set

$$G.x = \{g.x : g \in G\}.$$

Before turning to examples let us observe one small geometric point. Let x_1, x_2 be points of M lying on the same orbit, so there exists a $g \in G$ with $x_2 = g.x_1$. Observe that the mapping $M \to M$ defined by $x \to g.x$ is a transformation of M, (i.e. a bijection of M onto itself) which preserves orbits - by which we mean that a point is always mapped to another point lying on the same orbit - and which maps x_1 to x_2. What this amounts to is that one point on an orbit looks like any other; it is the property known as <u>homogeneity</u> of an orbit. The reader is urged to bear this notion in mind as a guiding intuition.

It is not the purpose of this chapter to pursue the general theory of groups acting on sets. Rather, we wish to concentrate the reader's attention on a class of geometric examples relevant to the mathematics of the next two chapters, and indeed providing genuine finite-dimensional analogues of the situations there studied. To this end

§2. Some Geometry of Jets

We start by recalling that in Chapter I we introduced a relation of equivalence on germs, so in particular on the set of all germs $(\mathbb{R}^n, 0) \to (\mathbb{R}^p, 0)$; indeed two such germs f, g are equivalent when there exist invertible germs h, k for which

$$f \circ h = k \circ g . \tag{1}$$

One can relativize this definition to d-jets as follows. f, g are said to have <u>equivalent d-jets</u> when there exist invertible germs h, k for which

$$j^d(f \circ h) = j^d(k \circ g). \tag{2}$$

In particular we can take f, g to be (germs at 0 of) d-jets, so inducing an equivalence relation on the jet-space $J^d(n, p)$. The geometric examples we have in mind arise from studying this equivalence relation not on the whole jet-space $J^d(n, p)$ but rather on the vector subspace $H^d(n, p)$ of all mappings $\mathbb{R}^n \to \mathbb{R}^p$ each of whose components (relative to the standard co-ordinates on $\mathbb{R}^p$) is a homogeneous polynomial of degree d in the standard co-ordinates $x_1, \ldots, x_n$ in $\mathbb{R}^n$. The geometry has its genesis in the following elementary, yet crucial, observation.

(2.1) <u>Two d-jets</u> f, g <u>in</u> $H^d(n, p)$ <u>are equivalent if and only if there exist invertible linear mappings</u> H, K <u>for which</u> $f \circ H = K \circ g$.

<u>Proof</u> The condition is certainly sufficient. To establish necessity, suppose there exist invertible germs h, k for which (2) holds. Write $\hat{\phi}$ for the Taylor series of a germ $\phi : (\mathbb{R}^n, 0) \to (\mathbb{R}^p, 0)$: then we have

$$\hat{h} = H + \text{terms of degree} \geq 2$$
$$\hat{k} = K + \text{terms of degree} \geq 2$$

with H, K invertible linear mappings. Thus, bearing in mind that the components of f, g are homogeneous polynomials of degree d, one has

$$(f \hat{\circ} h) = f \circ H + \text{terms of degree} > d$$
$$(k \hat{\circ} g) = K \circ g + \text{terms of degree} > d$$

so
$$f \circ H = j^d(f \circ h) = j^d(k \circ g) = K \circ g. \qquad \square$$

We can re-phrase the above as follows. Let $GL(s)$ denote the <u>general linear group</u> of all invertible linear mappings $\mathbb{R}^s \to \mathbb{R}^s$ under the operation of composition. The reader will readily check that we have an action of the group $GL(n) \times GL(p)$ on the vector space $H^d(n, p)$ given by $(H, K).f = K \circ f \circ H^{-1}$: and the condition for two d-jets f, g in

$H^d(n, p)$ to lie in the same orbit under this action is that there exist invertible linear mappings H, K for which $(H, K).g = f$, i.e. $f \circ H = K \circ g$. It follows immediately from (2.1) that the orbits in $H^d(n, p)$ under the action of $GL(n) \times GL(p)$ are precisely the equivalence classes described above. The next step in our study is to indicate the connexions with geometry. We start with

The Case $p = 1$

In this case an element of $H^d(n, 1)$ is just a homogeneous polynomial of degree d in $x_1, \ldots, x_n$ so (providing it is not zero) will define a hypersurface of degree d in real projective space $P\mathbb{R}^n$ of dimension $(n-1)$: for instance when $n = 2$ we are dealing with d points on a projective line, when $n = 3$ with a curve of degree d in the projective plane, and so on. And two elements of $H^d(n, 1)$ will lie in the same orbit if and only if the corresponding hypersurfaces can be obtained from each other by an invertible linear change of co-ordinates, i.e. are projectively equivalent. The difficulty of listing the orbits increases sharply with d: we shall allow ourselves the luxury of describing in detail some of the simpler cases relevant to the mathematics of the next two chapters.

One starts with $d = 1$. $H^1(n, 1)$ is the vector space of linear forms in n variables $x_1, \ldots, x_n$ and elementary linear algebra tells us that there are just two orbits namely that containing the zero form, and that containing all the non-zero forms.

The next case is $d = 2$. $H^2(n, 1)$ is the vector space of all quadratic forms in n variables $x_1, \ldots, x_n$ and elementary quadratic algebra tells us that any such form can be brought into the shape
$x_1^2 + \ldots + x_s^2 - x_{s+1}^2 - \ldots - x_r^2$ by an appropriate change of co-ordinates.

The numbers r, s are called the rank, index of the form, and (by Sylvester's Law of Inertia) are invariant under co-ordinate changes. However, if we multiply the form by -1 the rank remains invariant, though the index may change, so we work instead with the semi-index s' = min(s, r - s) of the form. Thus quadratic forms in $x_1, \ldots, x_n$ are classified by rank and semi-index. To make this even more explicit take the case when n = 2. A quadratic form in two variables x, y can be written $ax^2 + 2bxy + cy^2$ so can be identified with the point (a, b, c) in $\mathbb{R}^3$. We obtain four orbits as indicated in the picture below. The cone $b^2 = ac$ comprises the forms of rank 1 (the parabolic type) with the origin representing the zero form of rank 0 (the symbolic type). The remainder of the space comprises the forms of rank 2: indeed the inside of the cone corresponds to forms of semi-index 0 (the elliptic type) and the outside of the cone to forms of semi-index 1 (the hyperbolic type).

Let us continue with the case d = 3. $H^3(n, 1)$ is the vector space of all cubic forms in n variables $x_1, \ldots, x_n$. This case differs from the preceding ones in that a complete list of orbits is known only for $n \leq 4$. The first non-trivial case is n = 2, and since this case will arise naturally in the next chapter we shall take the opportunity to describe a once familiar bit of pure mathematics, the study of binary cubic forms

$$f(x, y) = \alpha x^3 + 3\beta x^2 y + 3\gamma xy^2 + \delta y^3.$$

It is easy to list the orbits using a little algebra. Let us (temporarily) allow the variables x, y to be complex and restrict ourselves to non-zero forms. Recall that a non-zero complex homogeneous polynomial of degree d in two variables factorizes into d linear factors (possibly with repetitions) in particular

$$f(x, y) = (a_1 x + b_1 y)(a_2 x + b_2 y)(a_3 x + b_3 y).$$

It follows that the zero set of f, i.e. the subset of $\mathbb{C}^2$ defined by the equation $f(x, y) = 0$ will comprise three lines through the origin. We shall distinguish four possible types of binary cubic form, corresponding to the following four types of triples. (The reason for the terminology will be explained shortly.)

  elliptic - all distinct and real
  hyperbolic - all distinct; one real and two complex
  parabolic - two distinct; both real, one repeated twice
  symbolic - one real line repeated thrice.

In fact these types are precisely the orbits we seek. Certainly two binary cubic forms of the same type will lie in the same orbit, since given two triples of lines through the origin (of the same type) we can always find a non-singular real linear mapping of $\mathbb{C}^2$ which maps the lines of the one triple to those of the other. And conversely two binary cubic forms in the same orbit are necessarily of the same type, since non-singular real linear mappings of $\mathbb{C}^2$ preserve the above types of triples of lines. We can obtain normal forms for non-zero binary cubic forms by just choosing an example of each of the four types just described. The standard choices are those given in the following table.

type	normal form
elliptic	$x^3 - xy^2$
hyperbolic	$x^3 + xy^2$
parabolic	$x^2 y$
symbolic	x^3

The reason for the terminology is as follows. Given a binary cubic form f one can associate with it a binary quadratic form H_f called the <u>Hessian</u>: it is defined to be

$$H_f = \frac{1}{36} \begin{vmatrix} \frac{\partial^2 f}{\partial x^2} & \frac{\partial^2 f}{\partial x \partial y} \\ \frac{\partial^2 f}{\partial y \partial x} & \frac{\partial^2 f}{\partial y^2} \end{vmatrix}$$

and a little arithmetic will verify that it is given by the formula

$$(\alpha\gamma - \beta^2)x^2 + (\alpha\delta - \beta\gamma)xy + (\beta\delta - \gamma^2)y^2 .$$

It is now an easy matter to verify that a binary cubic form f is elliptic, hyperbolic, parabolic or symbolic exactly according as the binary quadratic form H_f is; and that is the reason these words are used to describe the types of binary cubic form.

Although it is more difficult, one can obtain a visualization of binary cubic forms, just as we did for binary quadratic forms. The space of binary cubic forms can be identified with $\mathbb{R}^4$ by identifying $\alpha x^3 + 3\beta x^2 y + 3\gamma xy^2 + \delta y^3$ with the point $(\alpha, \beta, \gamma, \delta)$: the partition of the non-zero forms into four types yields a partition of $\mathbb{R}^4 - \{0\}$ into four sets, and by projecting this appropriately into $\mathbb{R}^3$ (we shall not go into the details) one arrives at the following delightful picture, dubbed the <u>umbilic bracelet</u>. It is obtained by rotating a deltoid (the curve traced by a fixed point on a circle rolling inside another circle of three times its radius) about an axis with a twist of $2\pi/3$ for each full circle. In this picture the cusped edge of the bracelet corresponds to the symbolic forms, the rest of the

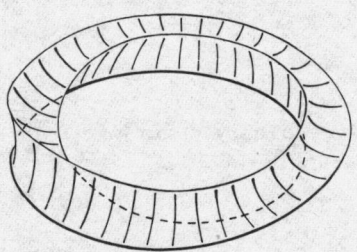

surface to the parabolic forms, the interior to the elliptic forms, and the exterior to the hyperbolic forms.

Let us pursue the possibility $d = 3$ a little further. The next case to consider is $n = 3$, i.e. ternary cubic forms in x, y, z: these are best thought of as cubic curves in the real projective plane $P\mathbb{R}^3$. The derivation of normal forms here is a lengthy exercise in the geometry of curves for details of which we refer the reader to texts on that subject; we shall merely quote the results. One starts with non-singular cubic curves.

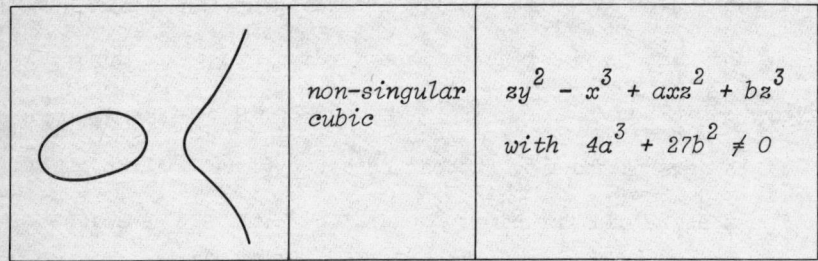

| | non-singular cubic | $zy^2 - x^3 + axz^2 + bz^3$ with $4a^3 + 27b^2 \neq 0$ |

To such a curve one associates the so-called j-invariant $j = 4a^3/4a^3 + 27b^2$: geometrically, this is the cross-ratio of the four lines through a point on the curve tangent to the curve elsewhere. It can be proved that two curves in this form lie in the same orbit if and only if the corresponding j's are equal, and the b's have the same sign: or if $b = 0$ then if and only if the a's have the same sign. Of course, a non-singular curve is automatically irreducible: the remaining irreducible curves are those which are singular, and these have exactly one real singular point. The tangents at the singular point can be distinct (the nodal case) or coincide (the cuspidal case). And in the nodal case one can made the finer distinction between the crunodal cubic (when the tangents are both real) and the acnodal cubic (when the tangents are complex conjugate). In this way one obtains three singular irreducible real cubic curves.

	crunodal cubic	$x^3 + y^2z + x^2z$
	acnodal cubic	$x^3 + y^2z - x^2z$
	cuspidal cubic	$x^3 - y^2z$

We are left with a motley array of reducible cubic curves. The cubic can reduce to a conic and a line, which may or may not be tangent to the conic; or it may reduce to three lines, giving rise to various possibilities.

	conic and line	$x(x^2 \pm y^2 \pm z^2)$
	conic and tangent	$y(xy - z^2)$
	triangle	$x(y^2 \pm z^2)$
	three concurrent lines	$x(x^2 \pm y^2)$
	two lines one repeated	$x^2 y$
	triple line	x^3

Rather than attempting to enlarge still further on the various situations which can arise by choosing special values of d, n we shall widen our class of examples by considering

The General Case $p \geq 2$

One thinks of an element of $H^d(n, p)$ as determining a vector subspace of $H^d(n, 1)$ of dimension $\leq p$, namely the subspace spanned by its components: for that reason elements of $H^d(n, p)$ can be called <u>linear systems in</u> $H^d(n, 1)$ <u>of dimension</u> $\leq p$. The simplest linear systems are those of dimension $\leq 2, 3$ which have long been known as <u>pencils</u>, <u>nets</u>, respectively, and whose study has whiled away the leisure hours of many a geometer in the past. We shall, reluctantly, confine ourselves to the simplest example of all, namely pencils of binary quadratic forms (i.e. d = 2, n = 2, p = 2) which we shall have occasion to recall in Chapter V.

Let us spell out one or two points. A pencil of binary quadratic forms will be a pair (q_1, q_2) of quadratic forms in two variables x, y and is to be identified with the subspace of $H^2(2, 1)$ spanned by these forms: it will therefore be a plane through the origin, a line through the origin, or just the origin itself. Recall that $H^2(2, 1)$ can be identified with $\mathbb{R}^3$ by identifying the quadratic form $ax^2 + 2bxy + cy^2$ with the point (a, b, c), and that the various types of binary quadratic form are separated exactly by the cone $b^2 = ac$. It is convenient to list the possible pencils by the way in which they lie relative to this cone. We claim that there are exactly seven possible pencils, as described in the following table. It should be clear to the reader first that any pencil of binary quadratic forms must be one of the seven types described, and second that these types are invariant under the action, i.e. if we apply an element of the group to a pencil of a given type the resulting pencil will be of the same type.

geometric description	normal form
plane intersecting the cone	$(xy, x^2 + y^2)$
plane lying outside the cone	$(xy, x^2 - y^2)$
plane tangent to the cone	(xy, x^2)
line inside the cone	$(x^2 + y^2, 0)$
line outside the cone	$(xy, 0)$
line tangent to the cone	$(x^2, 0)$
the origin	$(0, 0)$

What remains to be shown is that a pencil of a given type can be reduced to the normal form shown. The group acting here is $GL(2) \times GL(2)$. To simplify proofs we refer to elements of the first factor as <u>changes of co-ordinates</u>, and elements of the second factor as <u>changes of basis</u> in the pencil. It is easier to work backwards through the list. Suppose our pencil (q_1, q_2) represents a line through the origin; by a change of basis we can suppose $q_2 = 0$, and then a change of co-ordinates enables us to put q_1 into one of the forms $x^2 + y^2$, xy, x^2, according as it is elliptic, hyperbolic or parabolic, so yielding the three possible normal forms shown for the pencil. Suppose next that the pencil (q_1, q_2) represents a plane through the origin, tangent to the cone, so by a change of basis we can suppose q_2 singular, and then by a change of co-ordinates that $q_2 = x^2$: the plane being tangent to the cone q_1 cannot involve y^2, and a change of basis allows us to suppose $q_1 = xy$, as was required. Finally, suppose the pencil represents a plane through the origin not tangent to the cone. By a change of basis we can suppose q_1 hyperbolic, and then by a change of co-ordinates that $q_1 = xy$: subtracting off an appropriate multiple of this from q_2 we reduce q_2 to the form $\alpha x^2 + \beta y^2$ with $\alpha \neq 0$, $\beta \neq 0$ else the plane would be tangent to the cone. Obvious changes of co-ordinates and basis now bring the pencil to the form $(xy, x^2 \pm y^2)$, the + sign corresponding to the case when the plane intersects the cone, and the − sign to the case when it fails to intersect the cone. And that completes the derivation of the list.

§3. Smooth Actions of Lie Groups on Smooth Manifolds

We wish now to specialize the idea of an action Φ of a group G on a set M to the case when all three objects are "smooth", in a sense to be made precise. First, we consider the group. A __Lie group__ is a group G which is a smooth mnaifold, and for which multiplication $G \times G \to G$ and inversion $G \to G$ are smooth mappings.

__Example 1__ The general linear group $GL(n)$ is a Lie group. To see this it is convenient to identify it with the group of all non-singular real $n \times n$ matrices; it is therefore an open subset of the vector space $M(n)$ of all real $n \times n$ matrices, hence a smooth manifold. Further, multiplication of matrices in $M(n)$ is a polynomial mapping, so certainly smooth, and its restriction to matrices in $GL(n)$ will likewise be smooth. Finally, matrix inversion in $GL(n)$ is a rational mapping (with nowhere zero denominator) so smooth. Note incidentally that the tangent space to $GL(n)$ at any point can be identified with $M(n)$.

__Example 2__ The reader is left the task of checking that the cartesian product of two Lie groups is again a Lie group. In particular the product of two general linear groups will again be a Lie group.

By a <u>smooth action</u> of a Lie group G on a smooth manifold M we mean an action $\Phi : G \times M \to M$ for which Φ is smooth. The reader will readily check that the geometric examples given in the previous section are smooth actions of Lie groups on smooth manifolds. We intend to restrict ourselves entirely to the case when all the orbits are smooth submanifolds of M. In fact this turns out to be hardly any restriction at all, and it can be proved (though not too easily) that our geometric examples satisfy this condition. Further information on this point can be found in Appendix B. For instance, under the natural action of $GL(n) \times GL(p)$ on $H^1(n, p)$ the orbits provide the partition by rank of linear maps $\mathbb{R}^n \to \mathbb{R}^p$, and we saw in Chapter II that these are smooth manifolds.

In practice one needs to know how to compute the tangent space to an orbit at a point. The procedure is based on the following proposition.

(3.1) <u>Let $\Phi : G \times M \to M$ be a smooth action of a Lie group G on a smooth manifold M</u>. <u>It is assumed that all the orbits are smooth submanifolds of</u> M. <u>For any point</u> $x \in M$ <u>the natural mapping</u> $\Phi_x : G \to G.x$ <u>of the group onto the orbit given by</u> $g \to g.x$ <u>is a submersion</u>.

<u>Step 1</u> I claim that Φ_x has the same rank at every point in G. It suffices to show that the rank of Φ_x at any point $h \in G$ coincides with its rank at the identity element $1 \in G$. Let θ denote the diffeomorphism of G defined by $g \to hg$, and let Θ denote the diffeomorphism of M defined by $y \to h.y$. The commuting diagram of smooth mappings on the left then gives rise to the commuting diagram of differentials on the right. The vertical arrows in the diagram of differentials are linear isomorphisms, so $T_1 \Phi_x$ and $T_h \Phi_x$ have the same rank, as was claimed.

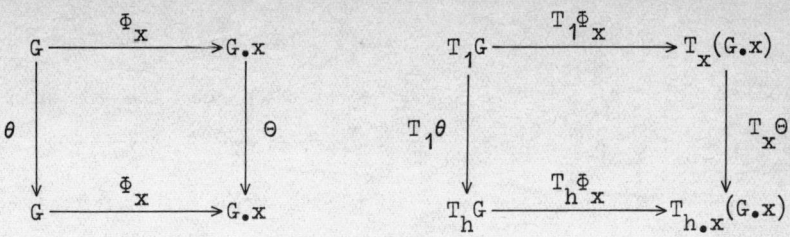

<u>Step 2</u> In view of Step 1 it suffices to show that Φ_x is submersive at some point in G, and that follows immediately from Sard's Theorem, guaranteeing the existence of at least one regular value, necessarily in the image of Φ_x. □

Thus the required tangent space $T_x(G.x)$ is the image under the differential $T_1\Phi_x$ of the tangent space T_1G to the group at the identity element. And that is how one goes about computing $T_x(G.x)$ in practice.

<u>Example 3</u> Let us return to our class of geometric examples namely the natural action of $GL(n) \times GL(p)$ on the vector space $H^d(n, p)$. Let $F : \mathbb{R}^n \to \mathbb{R}^p$ be an element of this vector space, with components $f_1, \ldots, f_p$ each a form of degree d in n variables $x_1, \ldots, x_n$. We shall compute the tangent space at F to the orbit through F. The natural mapping of the group onto the orbit is given by $(H, K) \to K \circ F \circ H^{-1}$, and we require the image of the differential of this mapping at the identity I, i.e. the vector sum of the images of the differentials at I of the mappings $H \to F \circ H^{-1}$ and $K \to K \circ F$. We consider these separately. Take first the mapping $H \to F \circ H^{-1}$. Observe first that it is the composite of matrix inversion $H \to H^{-1}$ with the mapping $H \to F \circ H$: since the differential of the former mapping (at any point) is invertible we see that the

images of the differentials at I of $H \to F \circ H^{-1}$, $H \to F \circ H$ will coincide so we need only consider the latter mapping. We compute the image of the differential as follows. Let $e_1, \ldots, e_n$ denote the standard basis vectors in $\mathbb{R}^n$, and for $1 \leq i, j \leq n$ let $\Delta_{ij} : \mathbb{R}^n \to \mathbb{R}^n$ be the linear mapping which maps e_j to e_i, and the remaining basis vectors to 0. Consider now the curves γ_{ij} in $GL(n)$ through I given by $\gamma_{ij}(t) = I + t\Delta_{ij}$ with t close to 0. Evidently the derivatives of these curves at I yield a basis for the vector space $M(n)$ of all linear mappings $\mathbb{R}^n \to \mathbb{R}^n$, which is the tangent space to $GL(n)$. It follows that the image of the differential at I of $H \to F \circ H$ will be spanned by the derivatives at 0 of the curves $F \circ \gamma_{ij}$, i.e. the derivatives at 0 with respect to t of $F(x_1, \ldots, x_j + tx_i, \ldots, x_n)$, which by the Chain Rule are the $x_i \frac{\partial F}{\partial x_j}$. Next, consider the mapping $K \to K \circ F$: this is the restriction of a linear mapping, so will coincide with its differential at I, and the image will be spanned by all $G = (g_1, \ldots, g_p)$ with just one g_i equal to an f_j, and all other g's zero.

Example 4   It is worthwhile isolating the case when $p = 1$. Let f be a form of degree d in n variables $x_1, \ldots, x_n$. The tangent space at f to the orbit through f will be the subspace of $H^d(n, 1)$ spanned by the $x_i \frac{\partial f}{\partial x_j}$ and f. In fact it is simpler than that. By Euler's Theorem $f = \frac{1}{d} \Sigma x_i \frac{\partial f}{\partial x_i}$, so lies in the subspace spanned by the $x_i \frac{\partial f}{\partial x_i}$: thus the required tangent space is just the subspace of $H^d(n, 1)$ spanned by the $x_i \frac{\partial f}{\partial x_j}$ with $1 \leq i, j \leq n$.

Example 5   By way of explicit illustration take the case of binary cubic forms, i.e. $d = 3$, $n = 2$, $p = 1$. And take $f = x^2 y$. The above tells us that the tangent space at f to the orbit through f will be spanned by

$x\frac{\partial f}{\partial x}$, $x\frac{\partial f}{\partial y}$, $y\frac{\partial f}{\partial x}$, $y\frac{\partial f}{\partial y}$ which are respectively the monomials $2x^2y$, $2xy^2$, x^3, x^2y, so a basis for the tangent space is given by x^2y, xy^2, x^3. It follows that the tangent space, hence the orbit, has dimension 3.

<u>Example 6</u> For an example involving linear systems consider pencils of binary quadratic forms, and in particular the pencil $F = (x^2, 0)$. By Example 3 the tangent space at F to the orbit through F will be spanned by $x\frac{\partial F}{\partial x}$, $y\frac{\partial F}{\partial x}$, $x\frac{\partial F}{\partial y}$, $y\frac{\partial F}{\partial y}$ together with $(x^2, 0)$, $(0, x^2)$; hence a basis for the tangent space is provided by $(x^2, 0)$, $(xy, 0)$, $(0, x^2)$ so the tangent space, and hence the orbit, is of dimension 3.

Let us return to the general situation of a smooth action $\Phi : G \times M \to M$ of a Lie group G on a smooth manifold M, supposing that all the orbits are smooth submanifolds of M. We can define the <u>codimension</u> cod x of any point $x \in M$ to be the codimension of the orbit $G.x$ in M. As we have seen, one of the broad problems is that of actually listing the orbits. In practice this tends to be a difficult, if not virtually impossible, question to answer. Experience suggests that the difficulty of listing the orbits of a given codimension may well increase with the value of the codimension: indeed, this is clear on an intuitive level, since the larger the codimension the more space there is in which the orbit can twist and turn, and hence the larger the number of possible types of behaviour. On this basis it should be most feasible to list those orbits of codimension 0.

This is worth re-phrasing. One calls $x \in M$ <u>stable</u> under the action of G when there exists a neighbourhood U of x in M such that every point $x' \in U$ lies in the same orbit as x: intuitively, this means that sufficiently small perturbations of x will not displace it from its orbit, hence the terminology. Note that the homogeneity property for orbits implies that if one point x on an orbit is stable then every point on that orbit is stable, and hence that the orbit is open in M. This geometric notion can be restated in algebraic terms. We call $x \in M$ <u>infinitesimally stable</u> when it has codimension zero. It is now a trivial exercise in linear algebra to see that x is stable if and only if it is infinitesimally stable, so that in the present context the distinction between the two notions is a fine one. However it is worth making, because in studying the singularities of smooth mappings one comes across analogous situations where both notions can be defined, but it is decidedly hard to show that they coincide.

<u>Example 7</u>   Consider again the natural action on binary cubic forms. Recall that there are five orbits, with representatives $0, x^3, x^2 y, x^3 \pm xy^2$; these have respective codimensions 4, 2, 1, 0 so we have two stable orbits, namely $x^3 \pm xy^2$.

<u>Example 8</u>   By way of contrast consider the natural action on ternary cubic forms, i.e. cubic curves. Here there are no stable orbits. One could prove this ad hoc by systematically computing the codimensions of the normal forms listed in §2. There is however a more illuminating argument. The group in question is $GL(3) \times GL(1)$. If we forget about the second factor here we obtain another action by the group $GL(3)$ alone. We claim that these two actions give rise to exactly the same orbits. What one has to show is that multiplication of a ternary cubic form by a scalar $\lambda \neq 0$ can be achieved by a change of co-ordinates: and this is clear since the change of co-ordinates $x \mapsto \lambda^{1/3} x, \; y \mapsto \lambda^{1/3} y, \; z \mapsto \lambda^{1/3} z$ achieves precisely that. Now the group $GL(3)$ has dimension 9, so the tangent space to any orbit at any point must be of dimension ≤ 9. On the other hand the space $H^3(3, 1)$ of ternary cubic form has dimension 10. It follows that any orbit has codimension ≥ 1.

Let us continue with the general situation of a smooth action $\Phi : G \times M \to M$. We wish to expand on the question of just how the action Φ behaves close to a point $x \in M$. The crude picture is that the orbits through points close to x flow smoothly together past x: and a very basic idea in studying this situation is to look at the way in which the orbits cut a (small) cross-section of this flow. (See Figure 1.)

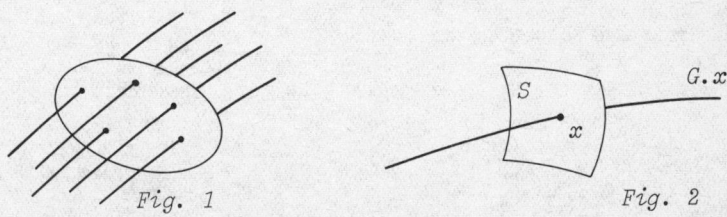

Fig. 1   Fig. 2

This crude idea can be made precise as follows. Consider a smooth submanifold S of M, with $x \in X$, which is transverse to the orbit $G.x$. (See Figure 2.) We shall certainly have $\dim M \leq \dim S + \dim(G.x)$: the least possible dimension for S will be when we have equality, i.e. when the dimension of S is cod x, and in that case we call S a <u>slice</u> at x. Under the diffeomorphism of M given by $m \mapsto g.m$, with $g \in G$ close to the identity, the slice S will slide slightly along the orbits: the basic picture which we wish to convey is that S will sweep out a small neighbourhood U of x in M. The formal mathematical statement of this reads as follows.

(3.1)   <u>Let</u> $\Phi : G \times M \to M$ <u>be a smooth action of a Lie group</u> G <u>on a smooth manifold</u> M, <u>and let</u> $x \in M$. <u>It is assumed that all the orbits are smooth submanifolds of</u> M. <u>The point</u> x <u>has a neighbourhood</u> U <u>with the following property: there is a smooth submanifold</u> H <u>of</u> G <u>containing the identity element, and a slice</u> S <u>at</u> x, <u>for which the restriction of</u> Φ <u>to</u> $H \times S$ <u>is a diffeomorphism onto</u> U.

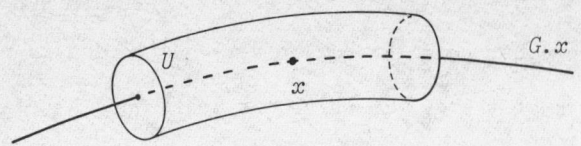

For obvious reasons such a neighbourhood U is said to have a <u>product structure</u>. One thinks of U as a small tube with the orbit G.x running down the middle.

<u>Step 1</u> We start by isolating a lemma which will be used twice in the main body of the proof. Suppose that B^b is a smooth submanifold of a smooth manifold A^a, and that $x \in B$: we claim that there exists a smooth submanifold C^c of A with $x \in C$ which is transverse to B and for which $a = b + c$. Indeed by (I.2.2) there is a diffeomorphism of an open neighbourhood of 0 in $\mathbb{R}^a$ onto a relatively open neighbourhood of x in A with 0 mapping to x, and $\mathbb{R}^b \times 0$ corresponding to B. If we put $c = a - b$ then $0 \times \mathbb{R}^c$ maps to a smooth submanifold C of A with the desired properties.

<u>Step 2</u> As we have already pointed out the natural mapping $\Phi_x : G \to G.x$ of the group onto the orbit is a submersion, so by (II.1.2) the inverse image G_x under this mapping of the point x will be a smooth submanifold of G of codimension the dimension of the orbit, with $1 \in G_x$. (Incidentally, G_x is a subgroup of G known as the <u>isotropy subgroup</u> at x.) By Step 1 there exists a smooth submanifold H of G with $1 \in H$ which is transverse to G_x with $\dim G = \dim G_x + \dim H$, i.e. H has the same dimension as the orbit G.x. Again by Step 1 there exists a slice S at x. Consider

now the restriction $\Psi : H \times S \to M$ of the action Φ. Observe that domain and target have the same dimension, and that the differential at $(1, x)$ is invertible. It follows from the Inverse Function Theorem that if H, S are small enough then Ψ maps $H \times S$ diffeomorphically onto a neighbourhood U of x in M. □

The reason for proving this result is to justify the intuitive idea that in order to study the action near x it will suffice to study the way in which the orbits cut a slice S. From a purely practical point of view this provides a considerable simplification since the dimension of the slice S may well be very small indeed compared to that of the manifold M. All this brings us to one of the central ideas of the subject, namely that of an "unfolding" of x, roughly speaking a parametrized slice at x.

§4. Transversal Unfoldings

Let $\Phi : G \times M \to M$ be a smooth action of a Lie group G on a smooth manifold M, and let $x \in M$. As usual, it is tacitly understood that all the orbits are smooth submanifolds of M. By an r-parameter unfolding of x we mean a germ $X : (\mathbb{R}^r, 0) \to (M, x) : X$ is said to be transverse when it is transverse to the orbit $G.x$ at x i.e.

$$T_0 X(\mathbb{R}^r) + T_x(G.x) = T_x M.$$

One pictures the situation something like this:

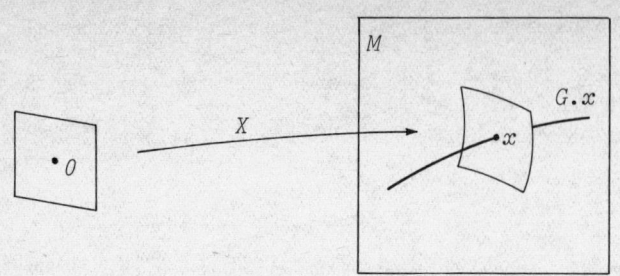

Now let x have codimension c. Given that X is a transverse unfolding it follows immediately that $r \geq c$: when $r = c$ we call X a __minimal__ transversal unfolding. The construction of an explicit minimal transversal unfolding X is a fairly straightforward matter. Let us restrict ourselves to the case when M is a linear space, so it is natural to look for linear unfoldings, i.e. those of the form

$$X(u_1, \ldots, u_c) = x + u_1 b_1 + \ldots u_c b_c$$

where $b_1, \ldots, b_c$ are fixed vectors in M. The condition for such an X to be transverse is that

$$\mathbb{R}\{b_1, \ldots, b_c\} + T_x(G.x) = M$$

where the first term represents the subspace spanned by $b_1, \ldots, b_c$, i.e. the image of the differential of X at 0. Thus all we need to do is to choose $b_1, \ldots, b_c$ to be a basis for a supplement of the tangent space $T_x(G.x)$.

__Example 1__ Consider the action of the product group $GL(2) \times GL(1)$ on the vector space $H^3(2, 1)$ of all binary cubic forms, discussed in §2. We shall compute transversal unfoldings for the non-zero normal forms. Recall that $H^3(2, 1)$ has as basis the monomials $x^3, x^2 y, xy^2, y^3$. We can compute the tangent space to the orbit at each normal form using the results of §3.

Bases for these tangent spaces can be found in the table below. It is a happy accident that in each case basis vectors for the tangent space can be chosen from the list of basis monomials, and the list of remaining monomials provides a basis for a supplement. Transversal unfoldings are then provided by the formula given above.

Normal Form	Basis for tangent space	Transversal Unfolding
$x^3 \pm xy^2$	x^3, x^2y, xy^2, y^3	$x^3 \pm xy^2$
x^2y	x^3, x^2y, xy^2	$x^2y + uy^3$
x^3	x^3, x^2y	$x^3 + uxy^2 + vy^3$

Let us pursue this example further to see just how the transversal unfolding can yield rather explicit information concerning the action close to a point. We shall describe the action close to x^3. As the point (u, v) moves over the plane $\mathbb{R}^2$ so the transversal unfolding $x^3 + uxy^2 + vy^3$ will cut all orbits through binary cubics close to x^3. We shall determine the values of u, v for which the unfolding represents a binary cubic of a given type. The associated Hessian quadratic form of the unfolding is $3ux^2 + 9vxy - u^2y^2$ with discriminant a multiple of $27v^2 + 4u^3$. The Hessian is the zero quadratic form if and only if $u = v = 0$: so the origin in the (u, v) plane is the only point which corresponds to a perfect cube. Let us now stay away from the origin. The unfolding will represent a parabolic binary cubic when the Hessian is parabolic, i.e. when the discriminant $v^2 + 4u^3 = 0$: this is the equation of a cuspidal cubic curve in the plane. (See the picture below.) At all other points in the plane $v^2 + 4u^3 \neq 0$. More precisely: inside the cuspidal cubic we have $v^2 + 4u^3 < 0$ corresponding to elliptic binary cubics, and outside the cuspidal cubic we have

83

$v^2 + 4u^3 > 0$ corresponding to hyperbolic binary cubics. Thus we arrive at the following picture, representing a cross-section of the flow of orbits close to x^3.

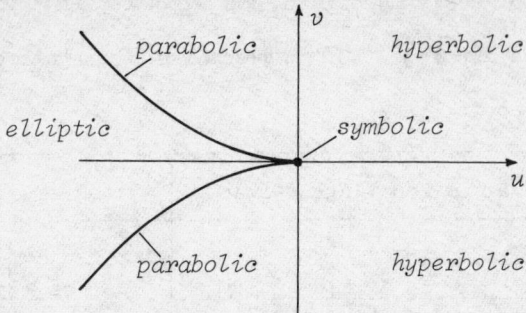

Some more interesting situations are provided by the transversal unfoldings of cubic curves.

Example 2   We shall study the transversal unfolding of the cubic curve $f = x(x^2 + yz)$, which is of the conic and chord type. (In fact in the real affine plane the conic is a parabola $y = -x^2$, and the chord is its axis $x = 0$.) A minor computation shows that the tangent space to the orbit at f is spanned by the monomials of degree 3 in x, y, z with y^3, z^3 deleted, so f has codimension 2 and a transversal unfolding is provided by

$$F = x(x^2 + yz) + uy^3 + vz^3.$$

As in the previous example we ask for the values of u, v for which F has a given type. Certainly, when $u = 0$, $v = 0$ we have the original conic and chord type. Suppose $u \neq 0$, $v = 0$: putting $\frac{\partial F}{\partial x}$, $\frac{\partial F}{\partial y}$, $\frac{\partial F}{\partial z}$ equal to zero one finds a singular point at $(0, 0, 1)$, and further computation shows that the tangent lines there are $x = 0$, $y = 0$ so one has a crunodal cubic. Likewise when $u = 0$, $v \neq 0$ one obtains a crunodal cubic. On the other hand when $u \neq 0$, $v \neq 0$ are sufficiently small the curve is non-singular, as a few lines of working will verify. Thus we obtain the following picture, representing a cross-section of the flow of orbits close to the conic and chord type.

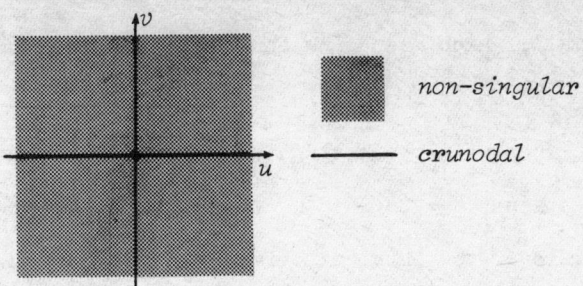

Example 3   Another pleasant illustration is provided by the transversal unfolding of the cubic curve $f = xyz$ of triangle type. In this case the tangent space to the orbit at f is spanned by the monomials of degree 3 in x, y, z with x^3, y^3, z^3 deleted, so f has codimension 3 and a transversal unfolding is provided by

$$F = xyz + ux^3 + vy^3 + wz^3.$$

We ask first for the values of u, v, w for which F is singular, i.e. there exist x, y, z not all zero for which

$$\frac{\partial F}{\partial x} = yz + 3ux^2 = 0$$

$$\frac{\partial F}{\partial y} = xz + 3vy^2 = 0$$

$$\frac{\partial F}{\partial z} = xy + 3wz^2 = 0.$$

Notice that if $u = 0$, or $v = 0$, or $w = 0$ then certainly this condition is satisfied. Let us see what happens when $u \neq 0$, $v \neq 0$, $w \neq 0$. Observe that if <u>one</u> of x, y, z vanishes then <u>all</u> of them vanish, so we can suppose $x \neq 0$, $y \neq 0$, $z \neq 0$. But then our equations yield

$$(27uvw + 1)x^2y^2z^2 = 0.$$

We are only considering what happens very close to f, so we can suppose u, v, w to be very small, which means the expression within brackets is $\neq 0$, so $x^2y^2z^2 = 0$ - which is impossible. In other words if u, v, w are all $\neq 0$ and small then F is non-singular. Let us look in more detail at what happens in the plane $u = 0$ in (u, v, w) space. Of course when both v, w are zero we recover the original triangle: when just one of v, w is zero we have the conic and chord type; and when both v, w are $\neq 0$ one obtains a crunodal cubic. By symmetry one has similar situations in the plane $v = 0$, and the plane $w = 0$. The nett result is the following picture of a cross-section of the flow or orbits past a cubic curve of triangle type.

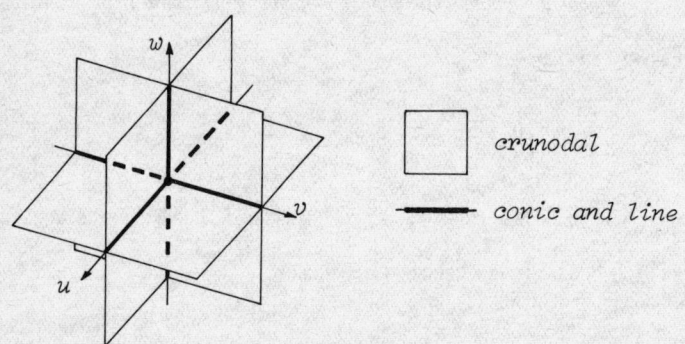

Example 4   For an example involving linear systems let us look at pencils of binary quadratic forms. We shall study the transversal unfolding of the pencil $F = (x^2, 0)$. The space of pencils is spanned by $(x^2, 0)$, $(xy, 0)$, $(y^2, 0)$, $(0, x^2)$, $(0, xy)$, $(0, y^2)$. In Example 6 of §3 we saw that the tangent space at F to the orbit through F is spanned by $(x^2, 0)$, $(xy, 0)$, $(0, x^2)$ so a basis for a supplement is provided by $(0, y^2)$, $(0, xy)$ and $(y^2, 0)$: thus F has codimension 3, and a transversal unfolding is given by $(x^2, 0) + u(0, y^2) + v(0, xy) + w(y^2, 0)$ i.e. $(x^2 + wy^2, uy^2 + vxy)$.

The objective now is to determine the values of u, v, w for which this pencil has one of the seven possible types listed in §2. Recall that we identified the space of binary quadratic forms with $\mathbb{R}^3$ by identifying $ax^2 + 2bxy + cy^2$ with (a, b, c), that the singular forms then correspond to the cone $b^2 = ac$, and that the types of pencil are separated by the way in which they lie relative to this singular cone. The first thing is to determine when the pencil represents a line through the origin: a simple computation shows that the binary quadratics $x^2 + wy^2$, $uy^2 + vxy$ are linearly dependent if and only if $u = 0$, $v = 0$, i.e. along the w-axis in (u, v, w) space. In that case when $w = 0$ we recover the original pencil F, a line tangent to the cone; when $w > 0$ we have a line inside the cone, and when $w < 0$ a line outside the cone. Now we ask what happens off the w-axis, when our pencil is a plane through the origin. To this end we consider how the plane cuts the cone. A typical element of the pencil is $\lambda(x^2 + wy^2) + \mu(uy^2 + vxy)$ i.e. $\lambda x^2 + \mu vxy + (\lambda w + \mu u)y^2$, which is singular when $(\mu v)^2 = 4\lambda(\lambda w + \mu w)$ i.e.

$$4w\lambda^2 + 4u\lambda\mu - v^2\mu^2 = 0. \qquad *$$

This expression is itself a binary quadratic in λ, μ with discriminant $16(u^2 + wv^2)$. The equation $u^2 + wv^2 = 0$ represents the vanishing of this discriminant, and is a surface in $\mathbb{R}^3$, called the <u>Whitney umbrella</u>. Note that it includes the whole w-axis, sometimes dubbed the <u>handle</u> of the umbrella. On this surface, but off the handle, the equation * has a unique ratio $\lambda : \mu$ for a solution so the pencil meets the cone $b^2 = ac$ in a line, so represents a plane tangent to the cone. Inside the umbrella one has $u^2 + wv^2 < 0$ so the equation * is satisfied by no real ratio $\lambda : \mu$, which means that the pencil represents a plane which does not intersect the cone.

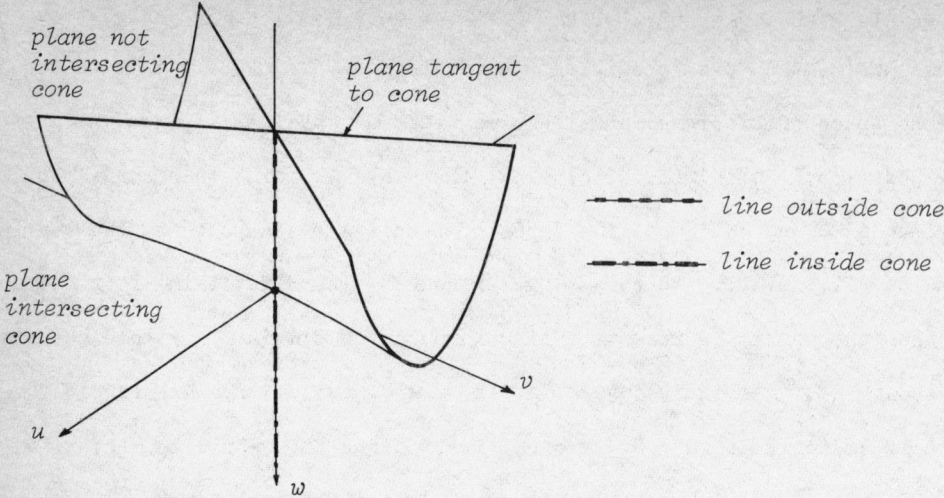

And outside the umbrella $u^2 + wv^2 > 0$, so the equation * is satisfied by two distinct real ratios $\lambda : \mu$ which means that the pencil represents a plane intersecting the cone in a pair of lines through the origin. Once again then we have obtained a simple picture representing a cross-section of the flow of orbits.

Of course, as we have presented them, all these examples provide little more than amusing exercises in elementary geometry. But there is more to it than that; there are deep questions in singularity theory, lying beyond the scope of this book, which one can only answer by going into the geometry of transversal unfoldings in considerable detail. Our objective was simply to lay bare the underlying ideas involved and to give the reader some feeling for the mechanics of the matter. Later in this book we shall meet analogous situations, lying just outside the present framework, where the idea of a transversal unfolding can be used to study the possible ways in which a germ of a smooth mapping can be deformed.

§5. Versal Unfoldings

We shall conclude our discussion of finite-dimensional unfoldings by expanding somewhat on the sense in which a transversal unfolding describes an action close to a point; this will provide us with a useful characterization of stable points in terms of their unfoldings. To this end we introduce a series of notions. As usual, Φ is a smooth action of a Lie group G on a smooth manifold M, and it is assumed that all the orbits are smooth submanifolds of M. We let $x \in M$.

Equivalence of Unfoldings

Two r-parameter unfoldings X_1, X_2 of x are said to be __equivalent__ when there exists an r-parameter unfolding $I : (\mathbb{R}^r, 0) \to (G, 1)$ of the identity element 1 in the group for which

$$X_2(u) = I(u) \cdot X_1(u).$$

Pictorially, the idea is that you can get from X_1 to X_2 by sliding smoothly down the orbits.

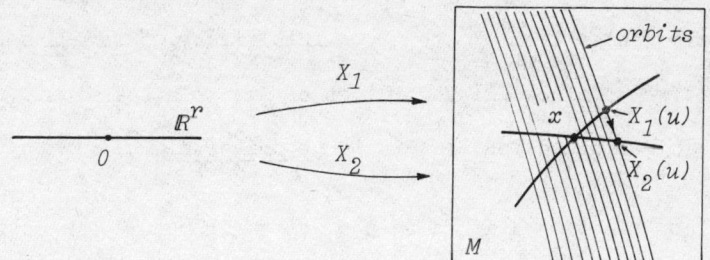

Induced Unfoldings

Suppose that X is an r-parameter unfolding of x, and that $H : (\mathbb{R}^s, 0) \to (\mathbb{R}^r, 0)$ is a germ. Then $Y = X \circ H$ is an s-parameter unfolding of x, said to be __induced by__ H. In this situation we refer to H

as a <u>change of parameter</u> and write $Y = H^*X$. One pictures the situation something like this.

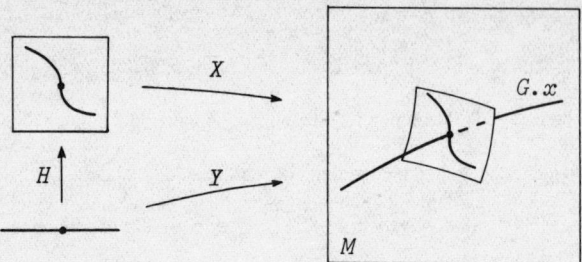

Morphisms of Unfoldings

Let X, Y be r, s-parameter unfoldings of x. <u>A morphism from</u> X <u>to</u> Y is a pair (H, I) with I an r-parameter unfolding of the identity 1, and with $H : (\mathbb{R}^s, 0) \to (\mathbb{R}^r, 0)$ a change of parameter, for which X is equivalent to the induced unfolding H^*Y via I. When $r = s$, and H is invertible we call the morphism an <u>isomorphism</u>.

Versal Unfoldings

An unfolding Y of x is said to be <u>versal</u> when for any unfolding X of x there exists a morphism from X to Y. Intuitively, this means that Y is sufficiently large to allow all unfoldings of x to appear in it. A versal unfolding of minimal dimension is said to be <u>universal</u>.

The basic fact about unfoldings is the following result connecting the algebraic notion of transversality with the geometric notion of versality.

<u>(5.1)</u>   <u>Let</u> Φ <u>be a smooth action of a Lie group</u> G <u>on a smooth manifold</u> M, <u>all of whose orbits are smooth submanifolds of</u> M. <u>A necessary and sufficient condition for an unfolding</u> $X : (\mathbb{R}^r, 0) \to (M, x)$ <u>to be versal is that</u>

it should be transversal.

Step 1   In which we establish necessity of the condition.   Suppose X versal. We have to show X is transversal, i.e. that

$$T_x M \subseteq T_0 X(\mathbb{R}^r) + T_x(G.x). \quad\text{———————(1)}$$

Consider any 1-parameter unfolding $Y : (\mathbb{R}, 0) \to (M, x)$. As X is versal there exists a morphism (H, I) from Y to X, i.e. $Y(v) = I(v).X(H(v))$. Take differentials to get

$$T_0 Y(\mathbb{R}) \subseteq T_0 Y(\mathbb{R}^r) + T_x(G.x). \quad\text{———————(2)}$$

(1) follows immediately from (2) since any tangent vector in $T_x M$ is a tangent vector to some curve through x, i.e. a 1-parameter unfolding Y.

Step 2   In which we establish sufficiency of the condition. Suppose X transversal. We wish to show X versal. Observe first that it will suffice to show that some unfolding induced by X is versal. Let us therefore replace X by an induced unfolding J^*X which is both transversal and minimal, and continue to denote it by the same letter X. (One can choose J to be a linear mapping into $\mathbb{R}^r$, with domain of the correct dimension, whose image is taken by $T_0 X$ to a supplement for $T_x(G.x)$ in $T_x M$.) Such a germ X is immersive, and the image of a sufficiently small representative will be a slice S at x. Recall now that x has a neighbourhood with a product structure, i.e. there exists a smooth submanifold H of G through the identity element 1 for which the germ Ψ at $(1, x)$ of the restriction of Φ to $H \times S$ is invertible. Consider now an arbitrary unfolding $Y : (\mathbb{R}^s, 0) \to (M, x)$ of x. We define germs J, Z by $J = \Pi_H \circ \Psi^{-1} \circ Y$, $Z = \Pi_S \circ \Psi^{-1} \circ Y$ where Π_H, Π_S are the germs at $(1, x)$ of the projections of $H \times S$ onto H, S respectively. It is immediate that $Y(u) = J(u).Z(u)$ so Y is equivalent to Z. On the other hand if we define the change of parameter $K : (\mathbb{R}^r, 0) \to (\mathbb{R}^r, 0)$ by $K = X^{-1} \circ Z$ we see that $Z = K^*X$ is an unfolding induced from X, and conclude that (K, J) is a morphism from Y to X. It follows that X is a versal unfolding. $\square$

An immediate consequence of this characterization is the following proposition justifying the use of the prefix in the term "universal".

(5.2) <u>Under the hypotheses of (5.1) any two universal unfoldings X, X' of x are isomorphic.</u>

<u>Proof</u> Suppose X, X' are r-parameter unfoldings. As X is versal, X' is equivalent to some induced unfolding H^*X. And as X' is universal H^*X is universal, hence is a minimal transversal unfolding by (5.1). That means that the image of the differential $T(X \circ H) = TX \circ TH$ is an r-dimensional subspace of T_xM, so the image of TH is likewise r-dimensional: it follows that TH is invertible, and hence H is invertible. The result follows. $\square$

Finally, as promised at the beginning of this section, we shall make use of these ideas to give a simple characterization of stable points. To this end let us call an r-parameter unfolding X of x <u>trivial</u> when it is equivalent to the r-parameter constant unfolding $(\mathbb{R}^r, 0) \to (M, x)$ given by $u \to x$. We then have

(5.3) <u>Under the hypotheses of (5.1) a necessary and sufficient condition for an element $x \in M$ to be stable is that every unfolding of x should be trivial</u>

<u>Necessity</u> Suppose x is stable, and that X is an r-parameter unfolding.

We have to show that X is equivalent to the r-parameter constant unfolding. Since x is stable the constant 0-parameter unfolding is transversal, hence versal by (5.1). Thus X is equivalent to an r-parameter unfolding induced from it, which is necessarily the constant r-parameter unfolding.

<u>Sufficiency</u> Conversely, suppose that every unfolding X of x is equivalent to a constant unfolding. In particular, take X to be a transversal unfolding. Certainly then some representative of X is a mapping into the orbit G.x, so X can only be transversal to the orbit when G.x has codimension 0, i.e. x is stable. □

The reader should be warned that the theory of this section is not particularly useful. Its virtue lies rather in the fact that it is an easily understood model for the ideas to be used in Chapter V.

IV Singular points of smooth functions

§1.   Some Basic Geometric Ideas

We come now to the meat of this book, the study of singular points of smooth mappings. In accordance with our philosophy of treating the simplest situations first we shall restrict this chapter to the case of singular points of smooth functions $f : N \to \mathbb{R}$ with N a smooth manifold. And we shall simplify life even further by studying germs of such functions under a rather finer notion of equivalence than that introduced in Chapter I: at first sight this may seem to be a complication rather than a simplification, but in fact it will enable us to finesse various algebraic difficulties. We call two germs of functions f_1, f_2 <u>right-equivalent</u> (or <u>$\mathcal{R}$-equivalent</u>) when there exists an invertible germ g for which the following diagram of germs commutes. Our broad intention

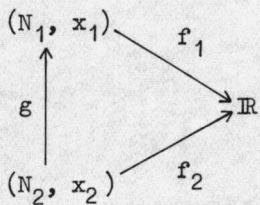

can now be formulated rather more accurately by saying that we wish to classify germs of functions under the relation of $\mathcal{R}$-equivalence. As such our programme is far too ambitious, but we shall see that we can gain some distance by using a little common sense. To avoid unnecessary symbolism we shall, throughout this chapter, adhere to the convention that <u>equivalence of two germs of functions is to mean $\mathcal{R}$-equivalence</u>: since we shall have no occasion to refer to the more general notion there should be no confusion on this point. And it will be useful to write $f_1 \sim f_2$ to mean that two germs f_1, f_2 of functions are equivalent.

The starting point is to try and set up our problem in such a way that it looks more like a problem which we know how to handle: the hope then is that we can proceed by analogy. That is the object of the present section. In setting up the basic geometric ideas we intend to be deliberately vague; the reason for this is that any systematic theory covering the situation we wish to study would lie far and away above the intended level of this book. All we shall do is to argue heuristically to arrive at precise interpretations of vague geometric ideas: we shall never use such arguments to prove propositions.

Our basic objects then are germs $g : (N, x) \to (\mathbb{R}, y)$. Taking a chart a at x we see that we can suppose $N = \mathbb{R}^n$, and $x = 0$. Thus the set of objects we wish to study is the set $\mathcal{E}_n$ of all germs $f : (\mathbb{R}^n, 0) \to (\mathbb{R}, y)$. Of course $\mathcal{E}_n$ is a real vector space under the natural operations of addition and scalar multiplication. In fact $\mathcal{E}_n$ has more algebraic structure than that; the operation of multiplication on $\mathbb{R}$ induces an operation of multiplication on $\mathcal{E}_n$, under which $\mathcal{E}_n$ becomes a real algebra. More information on the subject of real algebras can be found in Appendix C of this book. Next, let $\mathcal{R}_n$ denote the set of invertible germs $g : (\mathbb{R}^n, 0) \to (\mathbb{R}^n, 0)$, and observe that $\mathcal{R}_n$ is a group under the operation of composition. Now $\mathcal{R}_n$ acts on $\mathcal{E}_n$ by composition, i.e. we have an action $\mathcal{R}_n \times \mathcal{E}_n \to \mathcal{E}_n$ given by $(g, f) \mapsto f \circ g^{-1}$. Further, two germs in $\mathcal{E}_n$ will be equivalent if and only if they lie in the same orbit under this action. This observation, however trivial, is crucial in that it sets the scene in which we are to work.

However, pursuing the analogy with Chapter III is not quite that simple. What one would like to do is to introduce the "codimension" of a germ, and start the problem off by listing the orbits of fairly low codimension. The stumbling block here is that we have as yet only a group acting on a set, whereas we would like to have a Lie group acting smoothly on a smooth manifold. It is an unfortunate fact that $\mathcal{E}_n$ and $\mathcal{R}_n$ are not smooth manifolds in the sense in which that word has been used in this book. That need not be, and will not be, an insuperable difficulty. What we shall do is just to pretend that the action of $\mathcal{R}_n$ on $\mathcal{E}_n$ is a smooth action of a Lie group on a smooth manifold, and proceed by analogy.

This goes as follows. Choose a germ f in $\mathcal{E}_n$, and write just $\mathcal{R}.f$ for the orbit through f. We shall pretend that $\mathcal{R}.f$ is a "smooth submanifold" of $\mathcal{E}_n$, and look for a vector subspace of $\mathcal{E}_n$ which has some reasonable claim to the title of the "tangent space" to the orbit $\mathcal{R}.f$ at f. To do this we just mimic the theory of Chapter III. The natural mapping of the group onto the orbit is the mapping $\mathcal{R}_n \to \mathcal{E}_n$ given by $g \mapsto f \circ g^{-1}$, and one would expect the required "tangent space" to be the image of the "differential" at the identity 1 of this mapping. A preliminary simplification is to observe that $g \mapsto f \circ g^{-1}$ is the composite of $g \mapsto g^{-1}$ and $g \mapsto f \circ g$: one expects $g \mapsto g^{-1}$ to be a "diffeomorphism", so the required image should coincide with that of the "differential" at the identity of $g \to f \circ g$. Of course, one would expect this "differential" to be a linear mapping between vector spaces, and the first thing to get straight is just what these vector spaces should be. The domain should be the "tangent space" at the identity to the group $\mathcal{R}_n$. A little common sense will soon produce a candidate for this. Observe that $\mathcal{R}_n$ is a subset of the vector space $\mathcal{E}^0_{n,n}$ of all germs $g : (\mathbb{R}^n, 0) \to (\mathbb{R}^n, 0)$: moreover, we should really think of $\mathcal{R}_n$ as an "open" subset of $\mathcal{E}^0_{n,n}$ because if we disturb an invertible germ very slightly we still expect to get an invertible germ. On this basis we would certainly expect the "tangent space" to $\mathcal{R}_n$, at any point, to be just $\mathcal{E}^0_{n,n}$. The target of the "differential" should be the "tangent space" at x to $\mathcal{E}_n$, and since $\mathcal{E}_n$ is a vector space we would expect this to be just $\mathcal{E}_n$. Thus our "differential" should be a linear mapping $\mathcal{E}_{n,n}^0 \to \mathcal{E}_n$. The problem now is to find a sensible formula for the "differential". The line of attack to take here is suggested by Example 3 in §3 of Chapter III. We start with an arbitrary germ $g : (\mathbb{R}^n, 0) \to (\mathbb{R}^n, 0)$ i.e. an arbitrary "tangent vector" to $\mathcal{R}_n$ at 1: the components of g will be written $g_1, \ldots, g_n$. Now we can realize this germ as the "differential" at 0 of the curve through 1 in $\mathcal{R}_n$ given by $\gamma(t) = 1 + tg$ with t close to 0. The result of applying the sought for "differential" to g should then be the "differential" at 0 of the curve through f in $\mathcal{E}_n$ given by $t \to f(\gamma(t))$ i.e. the "differential" with respect to t at 0 of $f(x_1 + tg_1, \ldots, x_n + tg_n)$. Just doing this naively by the Chain Rule one gets the answer $g_1 \frac{\partial f}{\partial x_1} + \ldots g_n \frac{\partial f}{\partial x_n}$. To sum up: we expect the required "differential" to be the linear map $\mathcal{E}_{n,n}^0 \to \mathcal{E}_n$ given by the formula

$$(g_1, \ldots, g_n) \mapsto g_1 \frac{\partial f}{\partial x_1} + \ldots + g_n \frac{\partial f}{\partial x_n} . \qquad *$$

The required "tangent space" should be the image of this linear mapping, and this is neatly described in algebraic terms as follows. The germs $g_1, \ldots, g_n$ all have zero target, so belong to the ideal $\mathcal{M}_n$ in the algebra $\mathcal{E}_n$ comprising all germs with zero target. Let us denote by J_f the ideal in $\mathcal{E}_n$ generated by the partial derivatives $\frac{\partial f}{\partial x_1}, \ldots, \frac{\partial f}{\partial x_n}$, and call it the <u>Jacobian ideal</u> of f. The image of the linear mapping * above is just the product ideal $\mathcal{M}_n \cdot J_f$. In this way we come to the conclusion that a good candidate for the "tangent space" to the orbit f is $\mathcal{M}_n \cdot J_f$.

However, that is not quite the end of the story. We have described our model in the language of germs, but could just as well have worked with functions defined on a neighbourhood U of 0 in $\mathbb{R}^n$, and thought of the germ case as the limiting case when U is infinitesimally small. In that case we should replace $\mathcal{E}_n$ by the vector space $C^\infty(U, \mathbb{R})$ of all smooth functions defined on U. And we should replace the group $\mathcal{R}_n$ by the group Diff (U) of all diffeomorphisms U → U under the operation of composition. We would then have an action of Diff (U) on $C^\infty(U, \mathbb{R})$ given by the same formula as before. And the same heuristic reasoning as before would bring us to the conclusion that the required "tangent space" should be thought of as the image of a linear mapping given by the formula *. Note however one crucial difference. In the group $\mathcal{R}_n$ our germs are forced to preserve the origin, whilst in the group Diff (U) our diffeomorphisms need not have this property, no matter how small U is. All this indicates is that our model is oversimplified. We shall correct matters by allowing our germs $g_1, \ldots, g_n$ to have an arbitrary target, so to lie just in $\mathcal{E}_n$ rather than its ideal . The only difference this will make to the final answer is that the image of * will be simply the Jacobian ideal J_f.

On the basis of this heuristic reasoning we introduce the following formal definition. Let $f \in \mathcal{E}_n$. We define the tangent space to f to be the Jacobian ideal J_f. And we define the codimension cod f of f to be the codimension of the tangent space J_f in $\mathcal{E}_n$, i.e.

$$\operatorname{cod} f = \dim \mathcal{E}_n/J_f.$$

If cod f is finite we say that f is of <u>finite codimension</u>; otherwise, f is of <u>infinite codimension</u>. In practice we shall need to know how to decide whether or not a given germ f is of finite codimension, and if it is how to compute the codimension. There are some subtleties here which are worthy of explanation, and require a detailed discussion of the algebra $\mathcal{E}_n$: this then is the object of the next section.

§2. The Algebra $\mathcal{E}_n$

We have already observed that $\mathcal{E}_n$ is indeed a real algebra, in the technical sense of the word. (See Appendix C.) In this section we shall concern ourselves with purely algebraic matters so as to provide a convenient reference for succeeding sections. Observe first of all the following elementary fact.

(2.1) <u>A necessary and sufficient condition for a germ $f \in \mathcal{E}_n$ to be invertible (as an element of the ring $\mathcal{E}_n$) is that $f(0) \neq 0$.</u>

<u>Necessity</u> Suppose f is invertible, so we can find a germ $g \in \mathcal{E}_n$ with $fg = 1$: then $f(0)g(0) = 1$, so $f(0) \neq 0$.

<u>Sufficiency</u> Suppose $f \in \mathcal{E}_n$ satisfies $f(0) \neq 0$: then $g = 1/f$ is an inverse for f. □

This tells us that the ideal of all germs in $\mathcal{E}_n$ with target 0, has a rather special algebraic property.

(2.2) $\mathcal{M}_n$ <u>is the unique maximal ideal in</u> $\mathcal{E}_n$.

<u>Proof</u> Suppose $I \subseteq \mathcal{E}_n$ is an ideal with $\mathcal{M}_n \subset I \subseteq \mathcal{E}_n$, so we can find an $f \in I$ with $f(0) \neq 0$. f is invertible by (2.1) and hence $I = \mathcal{E}_n$. It follows that $\mathcal{M}_n$ is maximal. And the same argument establishes uniqueness. □

To some extent the importance of the ideal $\mathcal{M}_n$ lies in the fact that it allows a convenient algebraic description of certain ideals in $\mathcal{E}_n$ which frequently come into consideration. We have in mind the idea $\mathcal{I}_{k+1}$ of all

$f \in \mathcal{E}_n$ whose k-jet is zero (i.e. all partial derivatives of f of order $\leq k$ vanish at 0). As a preliminary we shall establish one extremely useful little fact – sometimes called the <u>Hadamard Lemma</u> – which we shall also have occasion to use in later chapters.

(2.3)   <u>Let U be a convex neighbourhood of 0 in $\mathbb{R}^n$, and let f be a smooth function defined on $U \times \mathbb{R}^q$ which vanishes on $0 \times \mathbb{R}^q$: there exist smooth functions $f_1, \ldots, f_n$ on $U \times \mathbb{R}^q$ with</u>

$$f = x_1 f_1 + \ldots + x_n f_n$$

<u>where $x_1, \ldots, x_n$ are the standard co-ordinate functions on $\mathbb{R}^n$.</u>

Proof

$$f(x_1, \ldots, x_n, y_1, \ldots, y_q) = \int_0^1 \frac{df}{dt}(tx_1, \ldots, tx_n, y_1, \ldots, y_q) dt$$

$$= \int_0^1 \sum_{i=1}^n x_i \frac{\partial f}{\partial x_i}(tx_1, \ldots, tx_n, y_1, \ldots, y_q) dt$$

$$= \sum_{i=1}^n x_i f_i(x_1, \ldots, x_n, y_1, \ldots, y_q)$$

if we take

$$f_i(x_1, \ldots, x_n, y_1, \ldots, y_q) = \int_0^1 \frac{\partial f}{\partial x_i}(tx_1, \ldots, tx_n, y_1, \ldots, y_q) dt.$$

□

Now we can characterise the ideal $\mathcal{I}_{k+1}$.

(2.4)   <u>$\mathcal{I}_k = \mathcal{M}_n^k$, and is generated by (the germs at 0 of) the monomials in $x_1, \ldots, x_n$ of degree $\leq k$.</u>

<u>Proof</u>   First, we establish $\mathcal{I}_k = \mathcal{M}_n^k$. The case $k = 1$ is trivial. It is clear that $\mathcal{M}_n^k \subseteq \mathcal{I}_k$, so it suffices to show $\mathcal{I}_k \subseteq \mathcal{M}_n^k$, which will follow by induction from $\mathcal{I}_k \subseteq \mathcal{M}_n \mathcal{I}_{k-1}$. This we prove using the Hadamard Lemma, in the special case $q = 0$. Indeed if $f \in \mathcal{I}_k$ then $f(0) = 0$, so that $f = x_1 f_1 + \ldots x_n f_n$, and it is clear from the construction of the f_i that they lie in $\mathcal{I}_{k-1}$: thus $f \in \mathcal{M}_n \mathcal{I}_{k-1}$. This argument shows, in particular, that $\mathcal{I}_1 = \mathcal{M}_n$ is generated by (the germs at 0 of) $x_1, \ldots, x_n$. It follows that $\mathcal{I}_k = \mathcal{M}_n^k$ is generated by (the germs at 0 of) the monomials of degree k in $x_1, \ldots, x_n$. □

In particular (2.4) shows that the ideals $\mathcal{I}_k$ are all finitely-generated. This is worth remarking on because the ring $\mathcal{E}_n$ is not Noetherian, i.e. not every ideal is finitely-generated. (We shall give an explicit example of such an ideal shortly.)

At this juncture it is worth saying something about the exact connexion between germs and their Taylor series. Of course a Taylor series is just a formal power series in several variables. Given a germ f in $\mathcal{E}_n$ we shall write its Taylor series as

$$\hat{f} = f(0) + \frac{1}{1!} \sum \frac{\partial f}{\partial x_i}(0) x_i + \frac{1}{2!} \sum \frac{\partial^2 f}{\partial x_i \partial x_j}(0) x_i x_j + \ldots$$

so $\hat{f}$ is in $\hat{\mathcal{E}}_n$, the algebra of formal real power series in n indeterminates $x_1, \ldots, x_n$. In this way we obtain a natural mapping $\mathcal{E}_n \to \hat{\mathcal{E}}_n$ given by $f \mapsto \hat{f}$, and one can check easily enough that it is a homomorphism of algebras. What is by no means so clear is the <u>Borel Lemma</u>.

(2.5)   <u>The algebra homomorphism $\mathcal{E}_n \to \hat{\mathcal{E}}_n$ given by $f \mapsto \hat{f}$ is surjective.</u>

The proof of the Borel Lemma is a slightly involved piece of analysis which we have isolated in Appendix D of this book so as not to interrupt the flow of relevant ideas. Notice that the kernel of the homomorphism $\mathscr{E}_n \to \hat{\mathscr{E}}_n$ is precisely $\mathscr{M}_n^\infty = \bigcap_{k=1}^\infty \mathscr{M}_n^k$ by (2.4). It follows that

$$\mathscr{E}_n / \mathscr{M}_n^\infty \cong \hat{\mathscr{E}}_n \, .$$

We can get a more finite version of this, as follows. Under the epimorphism $\mathscr{E}_n \to \hat{\mathscr{E}}_n$ the maximal ideal $\mathscr{M}_n$ in $\mathscr{E}_n$ maps to the maximal ideal $\hat{\mathscr{M}}_n$ in $\hat{\mathscr{E}}_n$ comprising all formal power series with zero constant term, and hence any power $\mathscr{M}_n^k$ maps to the power $\hat{\mathscr{M}}_n^k$. The nett result is that

$$\mathscr{E}_n / \mathscr{M}_n^k \cong \hat{\mathscr{E}}_n / \hat{\mathscr{M}}_n^k \, .$$

The particular virtue of this relation is that it tells us that the quotient space on the left is finite-dimensional. Indeed $\hat{\mathscr{M}}_n$ is generated by the indeterminates $x_1, \ldots, x_n$ so $\hat{\mathscr{M}}_n^k$ is generated by the monomials of degree k in $x_1, \ldots, x_n$ and the quotient space on the right can be identified with the real vector space of polynomials of degree $< k$ in $x_1, \ldots, x_n$, which is certainly of finite dimension.

In order to make further progress we shall require a rather pretty result from algebra, called the <u>Nakayama Lemma</u>.

(2.6) <u>Let $\mathscr{E}$ be a commutative ring with an identity element 1, and let $\mathscr{M}$ be an ideal in $\mathscr{E}$ with the property that $1 + x$ is invertible in $\mathscr{E}$ for any $x \in \mathscr{M}$. Further, let M be an $\mathscr{E}$-module, and let A, B be $\mathscr{E}$-submodules with A finitely generated. If $A \subseteq B + \mathscr{M} \cdot A$ then $A \subseteq B$.</u>

<u>Proof</u> Let $a_1, \ldots, a_t$ be generators for A. By hypothesis we can find $b_1, \ldots, b_t$ in B, and elements λ_{ij} in $\mathscr{M}$ for which for $1 \leq i \leq t$ we

can write

$$a_i = b_i + \lambda_{i1} \cdot a_1 + \ldots + \lambda_{it} \cdot a_t \qquad *$$

Introduce a $t \times t$ matrix over the ring $\mathcal{E}$ by $\Lambda = (\lambda_{ij})$ and take $a = (a_1, \ldots, a_t)$, $b = (b_1, \ldots, b_t)$ to be elements of the $\mathcal{E}$-module $M \times \ldots \times M$, with t factors. Then $*$ can be re-written as

$$(I - \Lambda)a = b$$

where I is the identity $t \times t$ matrix over the ring $\mathcal{E}$. It will suffice to show that $I - \Lambda$ is an invertible matrix, since then we can solve this system of linear equations for the $a_1, \ldots, a_t$ in terms of $b_1, \ldots, b_t$, which will show that $a_1, \ldots, a_t$ lie in B, as was required. To this end recall from linear algebra that a square matrix (over a commutative ring with an identity) is invertible if and only if its determinant is an invertible in the ring. So it suffices to show that $\det(I - \Lambda)$ is an invertible in $\mathcal{E}$. Observe that

$$\det(I - \Lambda) = 1 - \begin{pmatrix} \text{sum of products} \\ \text{of elements in } \mathcal{M} \end{pmatrix} = 1 - \lambda$$

say, with $\lambda \in \mathcal{M}$. And by hypothesis $1 - \lambda$ is an invertible in $\mathcal{E}$.

$\square$

In practice the rings $\mathcal{E}$ we have in mind are $\mathcal{E}_n$, $\hat{\mathcal{E}}_n$ and the ideals are $\mathcal{M}_n$, $\hat{\mathcal{M}}_n$ both of which satisfy the initial hypothesis of (2.6).

<u>Example 1</u>   The Nakayama Lemma allows a very simple proof that $\mathcal{E}_n$ is not a Noetherian ring. We take M to be the $\mathcal{E}_n$-module $\mathcal{E}_n$. And we take $A = \mathcal{M}_n^\infty = \bigcap_{k=1}^{\infty} \mathcal{M}_n^k$, B the trivial ideal. Clearly $A \subseteq B + \mathcal{M}_n \cdot A$. If $\mathcal{E}_n$ were Noetherian then A would be finitely generated, and the Nakayama Lemma would tell us that $A \subseteq B$, i.e. that A is trivial. However this

is false since there are standard examples in calculus of non-zero germs with zero Taylor series. It follows that $\mathcal{E}_n$ cannot be Noetherian.

Example 2 In $M = \mathcal{E}_2$ consider the ideals $A = \langle x^2, y^2 \rangle$ and $B = \langle x^2 + y^3, y^2 + x^3 \rangle$. Clearly $B \subseteq A$. We claim that indeed $A = B$, so need to show $A \subseteq B$. It is not immediately obvious to the eye that this is the case. Here the Nakayama Lemma provides an easy answer, because it tells us that it suffices to show $A \subseteq B + \mathcal{M}_2 \cdot A$. And this is clear as $\mathcal{M}_2 A$ is generated by x^3, xy^2, $x^2 y$, y^3.

Now we can return to the main theme. In the following $\mathcal{E}$ will denote either of $\mathcal{E}_n$, $\hat{\mathcal{E}}_n$ and $\mathcal{M}$ the corresponding maximal ideal $\mathcal{M}_n$, $\hat{\mathcal{M}}_n$. Let M be an $\mathcal{E}$-module, and let $I \subseteq M$ be an $\mathcal{E}$-submodule, so in particular a vector subspace of the real vector space M. We say that I has <u>finite codimension</u> in M when the quotient space M/I is finite-dimensional, i.e. the subspace I admits a finite-dimensional supplement in M. And in that case we define the <u>codimension</u> $\operatorname{cod} I$ of I in M to be the dimension of the quotient space M/I, i.e. the dimension of any supplement. The point of the next proposition is that it gives us a useful algebraic criterion for I to be of finite codimension in M.

(2.7) <u>Let M be an $\mathcal{E}$-module with a finite basis, and let $I \subseteq M$ be an $\mathcal{E}$-submodule. A necessary and sufficient condition for I to be of finite codimension in M is that there exists an integer $k \geq 1$ with $\mathcal{M}^k \cdot M \subseteq I$.</u>

Sufficiency Suppose there exists an integer $k \geq 1$ for which $\mathcal{M}^k \cdot M \subseteq I$. Certainly then

$$\dim M/I \leq \dim M/\mathcal{M}^k M.$$

Now I claim that $M/\mathcal{M}^k M$ is finite-dimensional. As M has a finite basis we can suppose $M = \mathcal{E}^s$ for some integer $s \geq 1$. Now $\mathcal{E}^s/\mathcal{M}^k.\mathcal{E}^s$ is naturally isomorphic with the product $\mathcal{E}/\mathcal{M}^k \times \ldots \times \mathcal{E}/\mathcal{M}^k$ with s factors, and since (as we have already seen) each factor in this product is finite-dimensional the product is as well. It follows that $\dim M/I < \infty$, so I has finite codimension in M.

<u>Necessity</u>   Suppose I is of finite codimension in M. Consider the descending sequence of $\mathcal{E}$-submodules

$$I + \mathcal{M}^0.M \supseteq I + \mathcal{M}^1.M \supseteq \ldots \supseteq I$$

Clearly, each strict inclusion in this sequence makes a contribution ≥ 1 to the codimension of I. Since the codimension is finite the inclusions, from some point onwards, must all be equalities. In particular there exists an integer $k \geq 1$ for which

$$I + \mathcal{M}^k.M = I + \mathcal{M}^{k+1}.M$$

which implies that $\mathcal{M}^k.M \subseteq I + \mathcal{M}.(\mathcal{M}^k.M)$. Since $\mathcal{M}^k.M$ is finitely-generated it follows from the Nakayama Lemma that $\mathcal{M}^k.M \subseteq I$, as required.

$\square$

Note one small point. As is clear from the final lines of the proof, the hypothesis $\mathcal{M}^k.M \subseteq I$ is equivalent to the apparently more complicated relation $\mathcal{M}^k.M \subseteq I + \mathcal{M}^{k+1}.M$. We chose the former relation for the statement of (2.7) because of its simplicity, but in practice it tends to be easier to verify the latter relation. A further point is that one can extract a bit more information from the proof of (2.7) than we have stated. Keeping to the same notation define

$$\text{cod}_k I = \dim \frac{I + \mathcal{M}^k.M}{I + \mathcal{M}^{k+1}.M}$$

for $k = 0, 1, 2, \ldots,$. The quotient space which appears here is necessarily of finite dimension, and its dimension is the codimension of $I + \mathscr{M}^{k+1}.M$ in $I + \mathscr{M}^k.M$.

(2.8) Under the hypotheses of (2.7) a necessary and sufficient condition for $I \subseteq M$ to be of finite codimension in M is that all but finitely many of the $\text{cod}_k I$ vanish: and in that case

$$\text{cod } I = \text{cod}_0 I + \text{cod}_1 I + \ldots .$$

<u>Necessity</u> Suppose I has finite codimension in M so by (2.7) there exists an integer $k \geq 1$ with $\mathscr{M}^k.M \subseteq I$. That implies $\text{cod}_j I = 0$ for all $j \geq k$.

<u>Sufficiency</u> If $\text{cod}_k I = 0$ then $I + \mathscr{M}^k.M = I + \mathscr{M}^{k+1}.M$ so $\mathscr{M}^k.M \subseteq I + \mathscr{M}^{k+1}.M$ which implies $\mathscr{M}^k.M \subseteq I$ by the Nakayama Lemma, i.e. I is of finite codimension in M.

The second statement in (2.8) follows immediately from the proof of (2.7). □

Next, let us expand a little on the kind of situation we have in mind. Write $\mathscr{E}_{n,p}$ for the real vector space of all germs $f : (\mathbb{R}^n, 0) \to (\mathbb{R}^p, y)$ with $y \in \mathbb{R}^p$. Such a germ has components $f_1, \ldots, f_p$ relative to the standard co-ordinates in $\mathbb{R}^n$, $\mathbb{R}^p$ each of which lies in $\mathscr{E}_n$. Thus we can think of $\mathscr{E}_{n,p} = \mathscr{E}_n \times \ldots \times \mathscr{E}_n$ (with p factors), and see that it is a natural example of an $\mathscr{E}_n$-module M with a finite basis. In practice we shall be thinking of I as the "tangent space" to a germ f: for instance in the case $p = 1$ we shall be thinking of $I = J_f$, the Jacobian ideal associated to a function-germ f. There is an important point to note here in connexion with the practicalities of computing the dimension of $\mathscr{E}_{n,p}/I$.

Let us write $\hat{\mathcal{E}}_{n,p} = \hat{\mathcal{E}}_n \times \ldots \times \hat{\mathcal{E}}_n$ (with p factors). The epimorphism $\mathcal{E}_n \to \hat{\mathcal{E}}_n$ of the Borel Lemma gives rise to a surjective linear mapping $\mathcal{E}_{n,p} \to \hat{\mathcal{E}}_{n,p}$ mapping I to $\hat{I}$, say. We claim the following.

(2.9) **A necessary and sufficient condition for I to be of finite codimension in $\mathcal{E}_{n,p}$ is that $\hat{I}$ should be of finite codimension in $\hat{\mathcal{E}}_{n,p}$; and in that case the two dimensions coincide.**

<u>Proof</u> Consider the composite of the surjective linear mappings $\mathcal{E}_{n,p} \to \hat{\mathcal{E}}_{n,p} \to \hat{\mathcal{E}}_{n,p}/\hat{I}$, the second being the natural projection onto the quotient. By previous work the kernel is $I + \mathcal{M}_n^\infty \cdot \mathcal{E}_{n,p}$ so that by elementary linear algebra

$$\frac{\mathcal{E}_{n,p}}{I + \mathcal{M}_n^\infty \cdot \mathcal{E}_{n,p}} \cong \frac{\hat{\mathcal{E}}_{n,p}}{\hat{I}} .$$

<u>Necessity</u> If I has finite codimension in $\mathcal{E}_{n,p}$ then so too does $I + \mathcal{M}_n^\infty \cdot \mathcal{E}_{n,p}$, and hence $\hat{I}$ has finite codimension in $\hat{\mathcal{E}}_{n,p}$ by the displayed isomorphism.

<u>Sufficiency</u> If $\hat{I}$ has finite codimension in $\hat{\mathcal{E}}_{n,p}$ then $I + \mathcal{M}_n^\infty \cdot \mathcal{E}_{n,p}$ has finite codimension in $\mathcal{E}_{n,p}$, by the displayed isomorphism. It follows from (2.7) that there exists an integer $k \geq 1$ with $\mathcal{M}_n^k \cdot \mathcal{E}_{n,p} \subseteq I + \mathcal{M}_n^\infty \cdot \mathcal{E}_{n,p} \subseteq I + \mathcal{M}_n^{k+1} \cdot \mathcal{E}_{n,p}$. The Nakayama Lemma then tells us that we have $\mathcal{M}_n^k \cdot \mathcal{E}_{n,p} \subseteq I$, and (2.7) that I has finite codimension in $\mathcal{E}_{n,p}$.

Finally, when I has finite codimension in $\mathcal{E}_{n,p}$ the relation $\mathcal{M}_n^k \cdot \mathcal{E}_{n,p} \subseteq I$ tells us that $\mathcal{M}_n^\infty \cdot \mathcal{E}_{n,p} \subseteq I$ so $I + \mathcal{M}_n^\infty \cdot \mathcal{E}_{n,p} = I$, and the displayed isomorphism reads

$$\mathcal{E}_{n,p}/I \cong \hat{\mathcal{E}}_{n,p}/\hat{I}$$

yielding the final statement of (2.9) □

The practical consequence of (2.9) is that when computing the codimension of I in $\mathcal{E}_{n,p}$ we can, and shall, replace all our germs by their Taylor series and handle them as formal power series, which are much easier to work with in practice. Bear this point in mind when you come to do such computations. Before resuming our session of algebra we shall digress to look at the sheer practicalities of just how one computes the codimension of a germ f in $\mathcal{E}_n$. Of course the first thing one wants to know is whether f has finite codimension: the next proposition at least gives one a very simple way of recognizing a germ of infinite codimension.

(2.10) <u>Suppose the germ f in $\mathcal{E}_n$ has finite positive codimension. Then the origin in $\mathbb{R}^n$ is an isolated singular point of any representative of f, i.e. there exists a neighbourhood of the origin in which the origin is the only singular point of the representative.</u>

<u>Proof</u> Observe first that $0 \in \mathbb{R}^n$ must indeed be a singular point of (any representative of) f. Indeed if some $\frac{\partial f}{\partial x_i}(0) \neq 0$ then $\frac{\partial f}{\partial x_i}$ would be an invertible element of $\mathcal{E}_n$, by (2.1) so $J_f = \mathcal{E}_n$, and f would have zero codimension. Since f has finite codimension we have $\mathscr{M}_n^k \subseteq J_f$ for some integer $k \geq 1$ by (2.6). In particular, this means that the monomials $x_1^k, \ldots, x_n^k$ can be written as linear combinations of $\frac{\partial f}{\partial x_1}, \ldots, \frac{\partial f}{\partial x_n}$ with coefficients in $\mathcal{E}_n$: at a singular point of f all these partial derivatives must vanish, so $x_1^k, \ldots, x_n^k$ must vanish as well, i.e. the singular point in question must be 0. □

<u>Example 3</u> Consider the germ $f(x, y, z) = y^2 - z^2x^2 + x^3$. Here the partial derivatives are

$$\frac{\partial f}{\partial x} = -2z^2 x + 3x^2 \; : \; \frac{\partial f}{\partial y} = 2y \; : \; \frac{\partial f}{\partial z} = -2zx^2$$

which vanish simultaneously precisely on the z-axis. Thus the origin in $\mathbb{R}^3$ is not an isolated singular point of f, and (2.10) tells us that f must be of infinite codimension. The set $f = 0$ is sketched below. Notice that the z-axis is precisely the line of "double points" where the surface intersects itself.

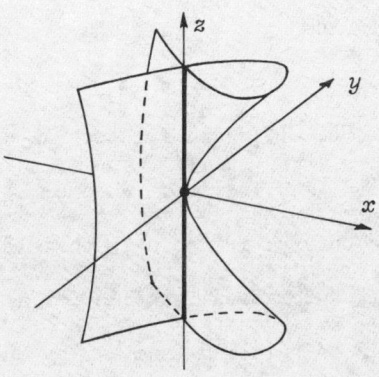

Thus (2.10) provides us with a simple method of showing that a germ is of infinite codimension. The strict converse of (2.10) is false, but there seems no harm in mentioning that in the <u>complex</u> case the converse does hold, though the proof requires considerable mathematical machinery. A good example is provided by the germ $f(x, y) = (x^2 + y^2)^2$: the reader will readily verify that as a <u>real</u> germ the origin is an isolated singular point, but that as a <u>complex</u> germ it is not, since any point on the lines $x + iy = 0$, $x - iy = 0$ is singular: thus the germ is of infinite codimension. Using these remarks the reader should find it a relatively straightforward matter to decide whether a given germ has finite codimension.

For the actual computation of the codimension we proceed as follows. Suppose that f is of finite codimension. The idea is to compute $\text{cod } f$ using (2.8). Let us abbreviate $\text{cod}_k J_f$ to $\text{cod}_k f$: we know that

$$\operatorname{cod} f = \operatorname{cod}_0 f + \operatorname{cod}_1 f + \dots$$

so we must successively compute $\operatorname{cod}_0 f$, $\operatorname{cod}_1 f$, ... until we reach a zero answer, and then add up the list of integers so obtained. The practicalities of the matter are worth expanding upon. To compute $\operatorname{cod}_k f$ we have to find a basis for a supplement of $J_f + \mathcal{M}^{k+1}$ in $J_f + \mathcal{M}^k$; clearly, this can be extracted from a list of monomials of degree k in the n variables $x_1, \dots, x_n$ which do not lie in $J_f + \mathcal{M}^{k+1}$. The first step in this computation is always trivial, as $\operatorname{cod}_0 f = \dim \mathcal{E}_n / \mathcal{M}_n = 1$. In practice then one takes successively $k = 1, 2, 3, \dots$; and for each k one has to decide, for each monomial of degree k in $x_1, \dots, x_n$, whether it lies in the ideal $J_f + \mathcal{M}^{k+1}$. A labour saving observation here is that if this condition holds for some monomial m then it automatically holds for all its <u>descendants</u>, i.e. the monomials which can be obtained from m by multiplying it be some other monomial. The reader is warned that the decision as to whether a given monomial lies in the relevant ideal may well involve quite a bit of work. A case which frequently appears in practice is $n = 2$. Here, the monomials in two variables x, y can be conveniently displayed as the following array

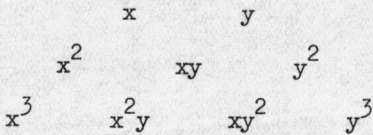

One works from the top of the array downwards. A simple way of recording the results of the computations is as follows. Suppose we have reached the kth row of the array, i.e. the monomials of degree k. Underline those monomials which appear in the relevant ideal $J_f + \mathcal{M}^{k+1}$: and then underline all its descendants in the array, to avoid unnecessary work. From the remaining monomials select a supplement for $J_f + \mathcal{M}^{k+1}$ in $J_f + \mathcal{M}^k$: the number of monomials selected is the integer $\operatorname{cod}_k f$. The computation finishes whenever one reaches a row of monomials which can all be underlined. The codimension is the sum of these $\operatorname{cod}_k f$, plus one — because we have to add $\operatorname{cod}_0 = 1$, assuming that f is a singular germ.

Example 4   Consider the germ $f(x) = x^{s+1}$ with s an integer ≥ 1. Here $J_f = \langle x^s \rangle$. For $0 \leq k \leq s - 1$ it is clear that the sole monomial of degree k in x, namely x^k, does not lie in the ideal $J_f + \mathcal{M}^{k+1}$: but x^s does. It follows immediately that $\operatorname{cod} f = s$.

Example 5   We shall compute the codimension of the germ $f(x, y) = x^2 y + y^4$, which will appear in §4 as the so-called "parabolic umbilic". Here $\frac{\partial f}{\partial x} = 2xy$, $\frac{\partial f}{\partial y} = x^2 + 4y^3$ vanish simultaneously (even in the complex case) only at the origin, so f should have finite codimension. We write down the array of monomials in x, y. (See Figure 1 below.) Neither of x, y lies in $J_f + \mathcal{M}^2$. Of x^2, xy, y^2 the first two certainly lie in $J_f + \mathcal{M}^+$ as $x^2 = \frac{\partial f}{\partial y} - 4y^3$, $xy = \frac{1}{2}\frac{\partial f}{\partial x}$, but y^2 does not; thus we underline x^2, xy and all their descendants in the array — namely x^3, x^2y, xy^2 in the third row, and x^4, x^3y, x^2y^2, xy^3 in the fourth row and so on. Of the monomials of degree 3 only y^3 need be considered, and it does not lie in $J_f + \mathcal{M}^4$. And of the monomials of degree 4 only y^4 need be considered, and it lies in $J_f + \mathcal{M}^5$ as $y^4 = \frac{1}{4}y\frac{\partial f}{\partial y} - \frac{1}{8}x\frac{\partial f}{\partial x}$. Thus all the monomials of degree 4 lie in $J_f + \mathcal{M}^4$, and the computation is finished. The codimension is the total number of monomials which have not be underlined (namely x, y, y^2, y^3) plus one, so $\operatorname{cod} f = 5$. We have, incidentally, shown that $\mathcal{M}^4 \subseteq J_f + \mathcal{M}^5$, and hence that $\mathcal{M}^4 \subseteq J_f$, by the Nakayama Lemma.

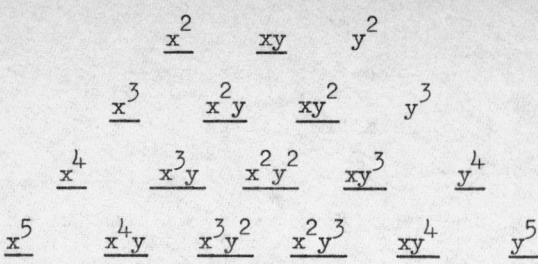

Figure 1

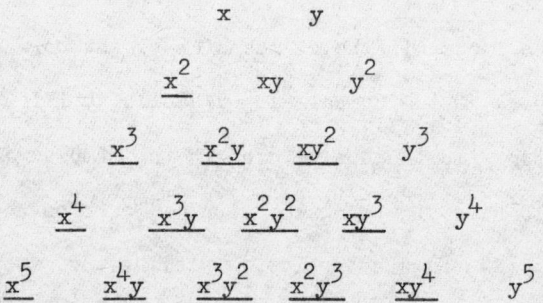

Figure 2

Example 6   The germ $f(x, y) = x^3 + xy^3$ is of finite codimension, indeed cod $f = 7$. The computation is very similar to that in Example 4 so we omit it. The array of monomials is illustrated in Figure 2.

Before going any further we should stop to verify that the codimension of a germ possesses the most basic property one would expect of it, namely that it is invariant under the relation of equivalence. To this end we introduce a new idea, which will be exploited later in this book. Given a germ $f : (\mathbb{R}^n, 0) \to (\mathbb{R}^p, 0)$ we obtain a mapping $f^* : \mathcal{E}_p \to \mathcal{E}_n$ by the formula $\lambda \to \lambda \circ f$. One checks easily that f^* is an algebra homomorphism, said to be <u>induced</u> by f. It is worth remarking that at the time of writing it seems to be an open problem whether every algebra homomorphism $\mathcal{E}_p \to \mathcal{E}_n$ has the form f^* for some germ $f : (\mathbb{R}^n, 0) \to (\mathbb{R}^p, 0)$. To the reader we leave the task of checking that the operation of taking induced homomorphisms has the following functorial properties,

(I) $(f \circ g)^* = g^* \circ f^*$.

(II) The germ at 0 of the identity map $\mathbb{R}^n \to \mathbb{R}^n$ induces the identity map $\mathcal{E}_n \to \mathcal{E}_n$, and is indeed the only such germ.

(III) It follows from (I) and (II) that if h is invertible then h^* is invertible and $(h^*)^{-1} = (h^{-1})^*$.

(2.11) Let $f : (\mathbb{R}^n, 0) \to (\mathbb{R}^n, 0)$ be a germ. A necessary and sufficient condition for the induced algebra homomorphism $f^* : \mathcal{E}_n \to \mathcal{E}_n$ to be an isomorphism is that f should be invertible.

Proof   Necessity of the condition is provided by III above. For sufficiency suppose f^* is an isomorphism, so maps the maximal ideal $\mathcal{M}_n$ onto itself. Indeed f^* will map the standard generators $x_1, \ldots, x_n$ for $\mathcal{M}_n$ to the components $f_1, \ldots, f_n$ of f, which must likewise be generators for $\mathcal{M}_n$. Consequently we can find elements λ_{ij} in $\mathcal{E}_n$ for which $1 \leq i \leq n$ we have

$$x_i = \lambda_{i1} f_1 + \ldots + \lambda_{in} f_n.$$

If we differentiate both sides of this relation with respect to x_j, and then evaluate at 0, we obtain

$$\delta_{ij} = \lambda_{i1}(0) \frac{\partial f_1}{\partial x_j}(0) + \ldots + \lambda_{in}(0) \frac{\partial f_n}{\partial x_j}(0)$$

where δ_{ij} denotes the Kronecker delta. Thus

$$I_n = \Lambda\left(\frac{\partial f_i}{\partial x_j}(0)\right)$$

where I_n denotes the identity $n \times n$ matrix, and Λ denotes the matrix of the λ_{ij}. It follows that the Jacobian matrix is invertible, and hence (from the Inverse Function Theorem) that f is invertible. $\square$

Now we are in a position to establish the invariance of the codimension, in the following precise sense.

(2.12) <u>If the germs f, g in $\mathcal{E}_n$ are equivalent then</u> cod f = cod g.

<u>Proof</u>   Since f, g are equivalent there exists an invertible germ $h : (\mathbb{R}^n, 0) \to (\mathbb{R}^n, 0)$ for which $g = f \circ h$. By (2.11) h induces an isomorphism $h^* : \mathcal{E}_n \to \mathcal{E}_n$ of algebras. We claim that $h^*(J_f) = J_g$: that will imply that the quotient spaces $\mathcal{E}_n/J_f$, $\mathcal{E}_n/J_g$ are isomorphic, from which the result follows. To establish the claim note that by the Chain Rule

$$\frac{\partial g}{\partial x_i} = \sum_{j=1}^{n} \left(\frac{\partial f}{\partial x_j} \circ h\right)\frac{\partial h_j}{\partial x_i} = \sum_{j=1}^{n} h^*\left(\frac{\partial f}{\partial x_j}\right)\frac{\partial h_j}{\partial x_i}$$

so $J_g \subseteq h^*(J_f)$. The same reasoning shows $J_f \subseteq (h^{-1})^* J_g = (h^*)^{-1} J_g$ yielding the reverse inclusion $h^*(J_f) \subseteq J_g$. $\square$

There is one final matter which we should clear up before passing to new things, namely the connexion between the codimensions of the ideals J_f, $\mathcal{M}J_f$. First of all,

(2.13) <u>The ideal $\mathcal{M}J_f$ has finite codimension if and only if J_f has finite codimension.</u>

Proof $\mathcal{M}J_f \subseteq J_f$, so certainly if $\mathcal{M}J_f$ has finite codimension then J_f has finite codimension. Conversely, suppose J_f has finite codimension, so $\mathcal{M}^k \subseteq J_f$ for some integer $k \geq 1$ by (2.7). But then $\mathcal{M}^{k+1} \subseteq \mathcal{M}J_f$ so, again by (2.7), we see that $\mathcal{M}J_f$ has finite codimension.

$\square$

(2.14) <u>Let $f \in \mathcal{E}_n$, and suppose that one (hence both) of $J_f, \mathcal{M}J_f$ has finite positive codimension:</u> then we have $\text{cod } \mathcal{M}J_f = n + \text{cod } J_f$.

Proof Observe first that J_f is the vector sum of the vector subspaces $\mathcal{M}J_f$ and $\mathbb{R}\{\frac{\partial f}{\partial x_1}, \ldots, \frac{\partial f}{\partial x_n}\}$, where the curly brackets indicate the vector subspace of $\mathcal{E}_n$ spanned by the first partial derivatives. Indeed an element of J_f has the form $u_1 \frac{\partial f}{\partial x_1} + \ldots + u_n \frac{\partial f}{\partial x_n}$ with the $u_i \in \mathcal{E}_n$; and the observation follows on writing $u_i(x) = (u_i(x) - u_i(0)) + u_i(0)$.

Next we claim that it is not possible to find germs $\xi_1, \ldots, \xi_n$ in $\mathcal{E}_n$, with at least one $\xi_i(0) \neq 0$, for which $\sum \xi_i \frac{\partial f}{\partial x_i} = 0$. To prove this it is convenient to confuse germs with their representatives. Think of $\xi_1, \ldots, \xi_n$ as the components of a vector field ξ on a neighbourhood of 0 in $\mathbb{R}^n$ with $\xi(0) \neq 0$. Recall now from (I.4.3) that the flow of ξ near 0 can be "straightened", i.e. we can find a diffeomorphism h of a neighbourhood of 0 in $\mathbb{R}^n$ with $h(0) = 0$ for which $\xi_i \circ h = \frac{\partial h_i}{\partial x_1}$, where $h_1, \ldots, h_n$ are the components of h. Observe now that $g = f \circ h$ is a germ in $\mathcal{E}_n$ equivalent to f, so likewise of finite positive codimension. Let us agree to write $b = h(a)$. Then $\frac{\partial g}{\partial x_1}(a) = \sum \xi_i(b) \frac{\partial f}{\partial x_i}(b) = 0$, so $\frac{\partial g}{\partial x_1}$ is identically zero on a neighbourhood of 0 in $\mathbb{R}^n$, i.e. g is independent of the variable x_1: by (2.10), the origin 0 is a singular point for g, so that every point near 0 on the x_1-axis is a singular point for g, and g has infinite codimension. This contradiction establishes our claim.

Now we can easily finish the proof. Taking $\xi_1, \ldots, \xi_n$ to be <u>real numbers</u> we see that $\frac{\partial f}{\partial x_1}, \ldots, \frac{\partial f}{\partial x_n}$ are linearly independent over the reals, and hence that the vector subspace $\mathbb{R}\left\{\frac{\partial f}{\partial x_1}, \ldots, \frac{\partial f}{\partial x_n}\right\}$ has dimension n. Further, it follows immediately from the above that the vector sum of $\mathcal{M} J_f$ and $\mathbb{R}\left\{\frac{\partial f}{\partial x_1}, \ldots, \frac{\partial f}{\partial x_n}\right\}$ is direct, i.e. that the intersection is the trivial subspace. The result now follows. $\square$

§3. Determinacy of Germs

Having completed our session of algebra we return to the main theme, namely the classification of germs in $\mathcal{E}_n$ of low codimension under $\mathcal{R}$-equivalence. The underlying philosophy is to reduce the problem to a finite-dimensional one. And the idea to do this is as follows. Take a germ $f \in \mathcal{E}_n$: f has a Taylor series

$$f(0) + \frac{1}{1!} \sum_{i=1}^{n} x_i \frac{\partial f}{\partial x_i}(0) + \frac{1}{2!} \sum_{i,j=1}^{n} x_i x_j \frac{\partial^2 f}{\partial x_i \partial x_j}(0) + \ldots$$

whose initial segments (the so-called Taylor polynomials of f) provide even better approximations (in some sense) to f. The hope is that a nice enough f will be equivalent to one of its Taylor polynomials, and then the problem of classification reduces to one in a <u>finite-dimensional</u> vector space of polynomials.

We introduce therefore the following definition. Let $k \geq 0$ be an integer. A germ $f \in \mathcal{E}_n$ is said to be <u>k-determined</u> when any germ $g \in \mathcal{E}_n$ with the same k-jet satisfies $f \sim g$: in other words a knowledge of all partial derivatives of order $\leq k$ completely determines the germ, up to equivalence. Note that the property of k-determinacy will be invariant under equivalence.

Example 1   Any non-singular germ $f \in \mathcal{E}_n$ is 1-determined. Indeed by (I.1.3) we know that f is equivalent to the germ $(x_1, \ldots, x_n) \to x_1$. And the same applies to any germ $g \in \mathcal{E}_n$ having the same 1-jet as f. Hence the result.

The basic result of this section is the following sufficient condition for k-determinacy.

(3.1)   Let $f \in \mathcal{E}_n$ be such that $\mathcal{M}^k \subseteq \mathcal{M} J_f$: then f is k-determined.

Proof   Let $g \in \mathcal{E}_n$ have the same k-jet as f. We have to prove $f \sim g$. Now f, g are points in the real vector space $\mathcal{E}_n$, so can be joined by a straight line. More formally, we define

$$F(x, t) = f_t(x) = (1 - t)f(x) + tg(x)$$

for $t \in \mathbb{R}$. Thus (f_t) is a 1-parameter family of germs with $f_0 = f$, $f_1 = g$. We intend to show that any two germs in this family are equivalent, which will prove the result. Since the real line is connected it will suffice to show that given any $s \in \mathbb{R}$ we have $f_t \sim f_s$ for t close to s. We claim that for this it will suffice to show that there exists a germ at $(0, s)$ of a smooth mapping $H : \mathbb{R}^n \times \mathbb{R} \to \mathbb{R}^n$ for which

(a) $H(x, s) = x$

(b) $H(0, t) = 0$

(c) $F\bigl(H(x, t), t\bigr) = F(x, s).$

To see this, write $h_t(x) = H(x, t)$. Now (a) tells us that h_s = identity. It follows that for t close to s the germ h_t must be invertible, since the determinant of the Jacobian matrix at 0 of h_t depends continuously on t. Further, the condition (b) ensures that h_t maps the origin in $\mathbb{R}^n$ to itself. Finally, the condition (c) can be re-written as $f_t \circ h_t = f_s$, so $f_t \sim f_s$ for t close to s, as was required. Notice incidentally that (c) is automatically satisfied for $t = s$, in view of (a). Thus it will be sufficient to replace (c) by the condition that the left hand side does not depend on t, i.e. has a zero derivative with respect to t. Written out in full this is the condition.

(c') $$\sum_{i=1}^{n} \frac{\partial H_i}{\partial t}(x, t) \frac{\partial F}{\partial x_i}\left(H(x, t), t\right) + \frac{\partial F}{\partial t}\left(H(x, t), t\right) = 0.$$

Our problem then is to construct a smooth mapping H which satisfies conditions (a), (b), (c'). We claim that it will suffice to construct a germ at $(0, s)$ of a smooth mapping $\xi : \mathbb{R}^n \times \mathbb{R} \to \mathbb{R}^n$ for which

(d) $$\sum \xi_i \frac{\partial F}{\partial x_i} = -\frac{\partial F}{\partial t}.$$

(e) $\xi_i(0, t) = 0.$

For suppose such a mapping ξ exists. We can think of ξ as a time-dependent vector field on $\mathbb{R}^n$. It has therefore a flow, i.e. a mapping H of the required type for which

(f) $$\frac{\partial H}{\partial t}(x, t) = \xi\left(H(x, t), t\right),$$

which can be supposed to satisfy the "initial condition" (a). A minor computation shows that (c') follows from (d) and (f) and that (b) follows from (a), (e) and (f).

It remains to establish the existence of a mapping ξ having the properties (d) and (e). The argument is algebraic, and does not depend on the particular value of s chosen: we shall therefore suppose $s = 0$. Also, since we wish to work simultaneously with functions of n variables $x_1, \ldots, x_n$ and functions of $(n+1)$ variables $x_1, \ldots, x_n, t$ we shall think of an element of $\mathcal{E}_n$ as an element of $\mathcal{E}_{n+1}$ which does not depend on t. And then condition (e) is simply the requirement that $\xi_1, \ldots, \xi_n$ should lie in $\mathcal{M}_n$. As a preliminary, note that

$$\frac{\partial F}{\partial t} = \frac{\partial}{\partial t}(1-t)f + tg = g - f \in \mathcal{M}_n^{k+1}$$

using the fact that $g - f$ has zero k-jet, and (2.4). Thus (d) will follow from

$$\mathcal{M}_n^{k+1} \subseteq \mathcal{M}_n \left\langle \frac{\partial F}{\partial x_1}, \ldots, \frac{\partial F}{\partial x_n} \right\rangle. \qquad *$$

Next we have

$$\mathcal{M}_n^{k+1} \subseteq \mathcal{M}_n^k \subseteq \mathcal{M}_n \left\langle \frac{\partial f}{\partial x_1}, \ldots, \frac{\partial f}{\partial x_n} \right\rangle$$

$$\subseteq \mathcal{M}_n \left\langle \frac{\partial F}{\partial x_1}, \ldots, \frac{\partial F}{\partial x_n} \right\rangle + \mathcal{M}_{n+1} \mathcal{M}_n^{k+1} \qquad **$$

the first inclusion being trivial, the second being the hypothesis of the theorem, and the third following from the fact that

$$\frac{\partial F}{\partial x_i} - \frac{\partial f}{\partial x_i} = \frac{t\partial(g-f)}{\partial x_i} \in \mathcal{M}_{n+1} \mathcal{M}_n^k$$

* now follows immediately from ** using the Nakayama Lemma, and the result is proved.

□

It is worth remarking that the condition (d) which appears in the above proof has a simple geometric interpretation. Recall that we took some germ f_t on the line joining f, g and set out to show that if we moved slightly along the line from f_t we remained in the orbit through f_t: thus we expected the direction of the line to be a "tangent vector" to the orbit through f_t. We think of the direction of the line being given by $\frac{\partial F}{\partial t}$. The relation (d) then says that $\frac{\partial F}{\partial t}$ lies in $\mathcal{M}J_{f_t}$, i.e. the "tangent space" to the orbit through f_t.

A word is also in order concerning the hypothesis of (3.1). The Nakayama Lemma tells us that the condition $\mathcal{M}^k \subseteq \mathcal{M}J_f$ is equivalent to the apparently more complicated condition $\mathcal{M}^k \subseteq \mathcal{M}J_f + \mathcal{M}^{k+1}$. We chose the former condition because it is rather simpler to remember though in practice it tends to be easier to establish the latter condition.

It is convenient to call $f \in \mathcal{E}_n$ <u>finitely-determined</u> when f is k-determined for some $k \geq 1$. It follows then from (2.7) and (3.1) that if f is of finite codimension then automatically f is finitely-determined. The practicalities of showing that a germ $f \in \mathcal{E}_n$ is k-determined are very similar to those involved in computing the codimension, discussed in §2. One checks successively whether $\mathcal{M}^k \subseteq \mathcal{M}J_f + \mathcal{M}^{k+1}$ for $k = 1, 2, \ldots$: and for a given k this is equivalent to showing that each monomial of degree k in $x_1, \ldots, x_n$ lies in the ideal $\mathcal{M}J_f + \mathcal{M}^{k+1}$.

<u>Example 1</u> Consider a germ $f \in \mathcal{E}_1$ for which the following conditions are satisfied.

$$f(0) = 0 : \frac{\partial f}{\partial x}(0) = 0 : \ldots : \frac{\partial^k f}{\partial x^k}(0) = 0 : \frac{\partial^{k+1} f}{\partial x^{k+1}}(0) \neq 0.$$

Clearly $f \in \mathcal{M}_1^k$, $f \notin \mathcal{M}_1^{k+1}$. It follows from the Hadamard Lemma that we can write $f = x^{k+1}g$ for some $g \in \mathcal{E}_1$ with $g(0) \neq 0$. Then $\frac{\partial f}{\partial x} = x^k \{(k+1)g + x\frac{\partial g}{\partial x}\}$: the expression in curly brackets is $\neq 0$ when $x = 0$, so invertible by (2.1), and one deduces that $J_f = \langle x^k \rangle = \mathcal{M}^k$ hence $\mathcal{M}J_f = \mathcal{M}^{k+1}$. (3.1) tells us that f is $(k+1)$-determined, so $f \sim cx^{k+1}$ where $c \neq 0$. An obvious change of co-ordinates then yields $f \sim \pm x^k$.

Example 2   The germ $f(x, y) = x^3 + y^3$ is 3-determined. Here $J_f = \langle x^2, y^2 \rangle$, so $\mathcal{M}J_f = \langle x^3, x^2 y, xy^2, y^3 \rangle = \mathcal{M}^3$, and the claim follows from (3.1)

Remark   Another sufficient condition for $f \in \mathcal{E}_n$ to be k-determined is that $\mathcal{M}^{k+1} \subseteq \mathcal{M}^2 J_f$. The proof of this is exactly the same as that of (3.1): one has simply to check that the last few lines of that proof go through equally well with this hypothesis. Of course, this is a slight improvement on (3.1) since $\mathcal{M}^{k+1} \subseteq \mathcal{M}^2 J_f$ is a weaker condition than $\mathcal{M}^k \subseteq \mathcal{M}J_f$.

Example 3   The germ $f(x, y) = x^4 + y^4$ is 4-determined. Here $J_f = x^3, y^3$ so $\mathcal{M}J_f = \langle x^4, x^3 y, xy^3, y^4 \rangle$ and the condition $\mathcal{M}^4 \subseteq \mathcal{M}J_f$ is <u>not</u> satisfied, so (3.1) does not allow us to conclude that f is 4-determined. On the other hand $\mathcal{M}^2 J_f = \langle x^5, x^4 y, x^3 y^2, x^2 y^3, xy^4, y^5 \rangle = \mathcal{M}^3$ so the improvement on (3.1) just mentioned allows us to conclude that f is indeed 4-determined.

Notice that if $f \in \mathcal{E}_n$ is k-determined then it is, in particular, equivalent to its kth Taylor polynomial. Thus the study of germs in $\mathcal{E}_n$ of finite codimension reduces to that of polynomial germs. In this connexion it is well to bear in mind that a polynomial germ of degree k which is finitely-determined is not necessarily k-determined. For instance the germ

121

$f(x, y) = x^5 + y^5$ is easily checked to be 6-determined, but can be shown not to be 5-determined.

§4. Classification of Germs of Codimension ≤ 5

We are now in a position to reap the rewards of our hard work in the previous section, and achieve our object of classifiing germs of low codimension. We start with germs of codimension 0.

(4.1) <u>A necessary and sufficient condition for a germ $f \in \mathcal{E}_n$ to have codimension 0 is that it be non-singular; and in that case it is equivalent to the germ $(x_1, \ldots, x_n) \to x_1$.</u>

<u>Necessity</u> Suppose f has codimension 0, so that $J_f = \mathcal{E}_n$ and the identity element 1 lies in J_f, i.e. we can find germs $\xi_1, \ldots, \xi_n$ in $\mathcal{E}_n$ for which $1 = \xi_1 \frac{\partial f}{\partial x_1} + \ldots + \xi_n \frac{\partial f}{\partial x_n}$. If we evaluate at 0 we see that at least one $\frac{\partial f}{\partial x_i}(0) \neq 0$, and hence f is non-singular.

<u>Sufficiency</u> Suppose f is non-singular, so at least one $\frac{\partial f}{\partial x_i}(0) \neq 0$. By (2.1) this means that $\frac{\partial f}{\partial x_i}$ is an invertible element of $\mathcal{E}_n$, and hence $J_f = \mathcal{E}_n$, so f has codimension 0.

Finally, a non-singular germ in $\mathcal{E}_n$ is equivalent to $(x_1, \ldots, x_n) \to x_1$ by (I.1.3). □

Next we consider germs of codimension 1. Remembering our discussion of functions in §5 of Chapter II we say that a germ $f \in \mathcal{M}_n^2$, i.e. a singular germ, is <u>non-degenerate</u> when the Hessian matrix

$$H_f = \left(\frac{\partial^2 f}{\partial x_i \partial x_j}(0)\right)$$

is non-singular. The next proposition, classifying germs of codimension 1, is a classic result in the calculus of variations, called the Morse Lemma.

(4.2)   <u>A necessary and sufficient condition for a germ $f \in \mathcal{M}_n^2$ to be of codimension 1 is that it be non-degenerate: and in that case f will be equivalent to a germ of the form</u> $x_1^2 + \ldots + x_s^2 - x_{s+1}^2 - \ldots - x_n^2$.

<u>Proof</u>   The main part of the proof consists of establishing that $f \in \mathcal{M}_n^2$ is non-degenerate if and only if $\mathcal{M} = J_f$. (Note: $\mathcal{M} \supseteq J_f$ automatically since the $\frac{\partial f}{\partial x_i}$ all vanish at 0.) To this end let Q denote the second polynomial of f - which is just a quadratic form in $x_1, \ldots, x_n$. Non-degeneracy of f is the same thing as non-degeneracy (in the normal sense) of the quadratic form Q, which in turn is equivalent to saying that the $\frac{\partial Q}{\partial x_i}$ span the vector space of all linear polynomials in $x_1, \ldots, x_n$ (by elementary linear algebra). Since we have $\frac{\partial Q}{\partial x_i} = \frac{\partial f}{\partial x_i}$ modulo $\mathcal{M}^2$, this last condition is equivalent to saying $\mathcal{M} = J_f + \mathcal{M}^2$, which in turn is equivalent to $\mathcal{M} = J_f$, using an obvious application of the Nakayama Lemma. The rest of the proof is now easy.

<u>Sufficiency</u>   Suppose f non-degenerate, so $J_f = \mathcal{M}$ by the above. Then codim f = dim $\mathcal{E}/J_f$ = dim $\mathcal{E}/\mathcal{M}$ = 1.

<u>Necessity</u>   Suppose f has codimension 1. Certainly then dim $\mathcal{E}/\mathcal{M}$ = 1. And if $J_f \subset \mathcal{M}$ then dim $\mathcal{E}/J_f > 1$, i.e. codim $f > 1$, which is impossible. Thus $J_f = \mathcal{M}$, and f is non-degenerate, by the above discussion.

Finally, we have to verify that f has the quoted normal form. As we saw above, non-degeneracy of f is equivalent to $\mathcal{M} = J_f$ which implies $\mathcal{M}^2 \subseteq \mathcal{M} J_f$. It follows immediately from (3.1) that f is 2-determined, so equivalent to its second Taylor polynomial Q. Recall now from quadratic algebra that Q, being a non-degenerate quadratic form, is equivalent (under a linear change of co-ordinates) to the normal form $x_1^2 + \ldots x_s^2 - x_{s+1}^2 - \ldots x_1^2 + \ldots x_s^2 - x_{s+1}^2 - \ldots - x_n^2$ where s is the index of Q. That concludes the proof. $\square$

The geometric meaning of the Morse Lemma is worth spelling out. Suppose we have a smooth function $f : \mathbb{R}^n \to \mathbb{R}$ with $f(0) = 0$ admitting a non-degenerate critical point at 0. The Morse Lemma tells us that there is a diffeomorphism of one neighbourhood of I in $\mathbb{R}^n$ onto another under which the fibres $f^{-1}(\epsilon)$ correspond to sets $x_1^2 + \ldots + x_s^2 - x_{s+1}^2 - \ldots , - x_n^2 = \epsilon$, for small enough ϵ. In particular, $f^{-1}(0)$ corresponds to a quadric cone; and for $\epsilon \neq 0$ the relevant part of $f^{-1}(\epsilon)$ is a smooth manifold diffeomorphic to part of a non-singular quadric surface. Here is a picture for the case $n = 2$, $s = 1$.

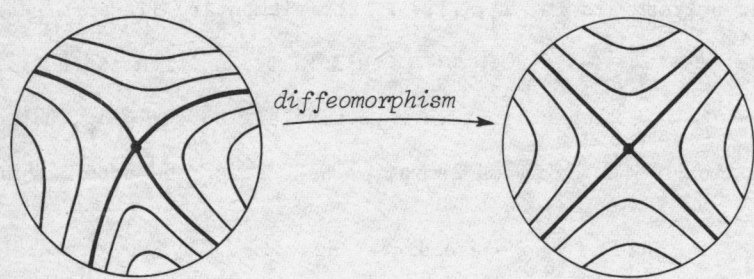

To make this even more convincing recall again our discussion of functions in §5 of Chapter II. There we showed that any function $f : \mathbb{R}^n \to \mathbb{R}$ could be approximated as closely as we please by one $f_t : \mathbb{R}^n \to \mathbb{R}$ which admits only non-degenerate critical points; and indeed that f_t could be obtained from f be a linear deformation. One of our examples was the monkey saddle

$$f(x, y) = x^3 - 3xy^2$$

which admits a degenerate critical point at 0. In this case we could take

$$f_t(x, y) = x^3 - 3xy^2 - tx$$

as the approximating function, for $t > 0$ as small as we please. The degenerate critical point of f at 0 "splits" into two non-degenerate critical points at the points $\left(\pm\sqrt{\frac{t}{3}}, 0\right)$. And the diagram of level curves $f_t^{-1}(\epsilon)$ is as below: the reader will note that close to the critical points we have precisely the situation described above.

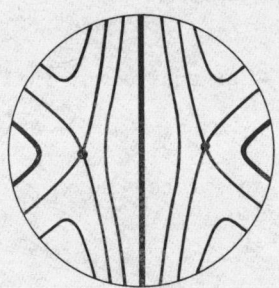

To push the classification further for germs of codimension ≥ 2 we shall require one more bit of machinery. Suppose that $f \in \mathscr{M}_n^2$ has codimension ≥ 2. Certainly then the Hessian matrix H_f is singular, so has rank $r < n$: the non-negative integer $c = n - r$ is called the <u>corank</u> of g. (One easily checks that it is invariant under equivalence.) The extra machinery which we require is the following result, known as the <u>Splitting Lemma</u>, and proved originally under less restrictive hypotheses.

(4.3)   <u>Let</u> $f \in \mathscr{M}_n^2$ <u>be a finitely-determined germ of corank</u> c; f <u>is equivalent to a germ</u>

$$g(x_1, \ldots, x_c) + \delta_{c+1} x_{c+1}^2 + \ldots + \delta_n x_n^2$$

where $g \in \mathcal{M}_c^3$ and each $\delta_i = \pm 1$.

Proof   We shall adopt the following notation. Given two germs ϕ, ψ in $\mathcal{E}_n$ we write $\phi \underset{k}{\sim} \psi$ when there exists an invertible germ $h : (\mathbb{R}^n, 0) \to (\mathbb{R}^n, 0)$ for which $\phi \circ k, \psi$ have the same k-jet. Observe that $\underset{k}{\sim}$ is an equivalence relation. We intend to show, by induction on k, that there exists a germ $g_k \in \mathcal{M}_c^3$, which is polynomial of degree $\leq k$, for which

$$f(x_1, \ldots, x_n) \underset{k}{\sim} g_k(x_1, \ldots, x_c) + \delta_{c+1} x_{c+1}^2 + \ldots + \delta_n x_n^2 . \qquad *$$

This will suffice to prove the result, since there exists a k for which f is equivalent to its k-jet.

The induction starts when $k = 2$. The 2-jet of f is a quadratic form in n variables of rank $r = n - c$, so equivalent (under a linear change of coordinates) to a quadratic form $\delta_{c+1} x_{c+1}^2 + \ldots + \delta_n x_n^2$ with each $\delta_i = \pm 1$, by standard quadratic algebra: and we can take $g_2 = 0$. For the induction step we assume *. Certainly then

$$f(x_1, \ldots, x_n) \underset{k+1}{\sim} g_k(x_1, \ldots, x_c) + \delta_{c+1} x_{c+1}^2 + \ldots + \delta_n x_n^2 + H(x_1, \ldots, x_n)$$
$$**$$

where H is a homogeneous polynomial of degree $(k + 1)$. Now write, as we evidently can do,

$$H(x_1, \ldots, x_n) = h(x_1, \ldots, x_c) + x_{c+1} H_{c+1}(x_1, \ldots, x_n) + \ldots$$
$$+ x_n H_n(x_1, \ldots, x_n)$$

where h is a homogeneous polynomial of degree $(k + 1)$ and $H_{c+1}, \ldots, H_n$ are homogeneous polynomials of degree k. We use these latter polynomials to define $\phi : (\mathbb{R}^n, 0) \to (\mathbb{R}^n, 0)$ by taking its components to be

$$\phi_1(x_1, \ldots, x_n) = x_1$$
$$\vdots$$
$$\phi_c(x_1, \ldots, x_n) = x_c$$
$$\phi_{c+1}(x_1, \ldots, x_n) = x_{c+1} - \frac{1}{2\delta_{c+1}} H_{c+1}(x_1, \ldots, x_n)$$
$$\vdots$$
$$\phi_n(x_1, \ldots, x_n) = x_n - \frac{1}{2\delta_n} H_n(x_1, \ldots, x_n).$$

Note that the Jacobian matrix of ϕ is the identity $n \times n$ matrix, so ϕ is invertible. Substituting $\phi_1, \ldots, \phi_n$ for $x_1, \ldots, x_n$ in ** we obtain

$$f(x_1, \ldots, x_n) \underset{k+1}{\sim} \tilde{g}_{k+1}(x_1, \ldots, x_c) + \delta_{c+1} x_{c+1}^2 + \ldots + \delta_n x_n^2$$

where $g_{k+1} = g_k + h$. That completes the induction step, and establishes the result. $\square$

It is worth remarking that the germs f, g which appear in the statement of the Splitting Lemma are necessarily of the same codimension. The proof can safely be left as an exercise for the reader. Our first application of the Splitting Lemma is to establish a broad connexion between the corank and the codimension of a germ, which will be crucial to the remainder of the classification. Roughly speaking, it is that as the corank increases so does the codimension, only rather more quickly. Precisely,

(4.4)   Let $f \in \mathscr{M}_n^2$ be a germ of finite codimension, and corank c: the codimension of f is $\geq \frac{c(c+1)}{2} + 1$.

<u>Proof</u>   We have already observed in §3 that a germ of finite codimension is automatically finitely-determined, so the Splitting Lemma applies, and tells us that f is equivalent to a germ

$$g(x_1, \ldots, x_c) \pm x_{c+1}^2 \pm \ldots \pm x_n^2$$

with $g \in \mathcal{M}_c^3$: and (vide the remark above) f has the same codimension as g. Note that $I = \mathcal{M}_c J_g \subseteq \mathcal{M}_c^3$ and that by (2.8)

$$\text{cod } I = \text{cod}_0 I + \text{cod}_1 I + \text{cod}_2 I + \text{cod}_3 I + \ldots .$$

But

$$\text{cod}_0 I = \dim \frac{I + \mathcal{E}_c}{I + \mathcal{M}_c} = \dim \frac{\mathcal{E}_c}{\mathcal{M}_c} = 1$$

$$\text{cod}_1 I = \dim \frac{I + \mathcal{M}_c}{I + \mathcal{M}_c^2} = \dim \frac{\mathcal{M}_c}{\mathcal{M}_c^2} = c$$

$$\text{cod}_2 I = \dim \frac{I + \mathcal{M}_c^2}{I + \mathcal{M}_c^3} = \dim \frac{\mathcal{M}_c^2}{\mathcal{M}_c^3} = \frac{c(c+1)}{2}$$

so

$$1 + c + \frac{c(c+1)}{2} \leq \text{cod } I = c + \text{cod } J_g$$

using (2.14), from which the result follows. $\square$

Thus, for instance, we have the following rough estimates

$$c = 1 \; : \; \text{cod } f \geq 2$$
$$c = 2 \; : \; \text{cod } f \geq 4$$
$$c = 3 \; : \; \text{cod } f \geq 7$$
$$\vdots \qquad \vdots$$

The consequence of (4.4) which we need to isolate is that <u>if</u> $\text{cod } f \leq 5$ <u>then corank</u> $f \leq 2$.   It will therefore suffice to look at germs of corank 1 and corank 2.   We are now in a position to state the remainder of the classification up to codimension 5: the result is due to R. Thom.

(4.5) Let $f \in \mathcal{M}_n^2$ have codimension ≥ 2 and ≤ 5: then (up to addition of a non-degenerate quadratic form in further variables, and multiplication by ± 1) f is equivalent to one of the seven germs on the following list.

corank	codim	germ	name
1	2	x^3	fold
	3	x^4	cusp
	4	x^5	swallowtail
	5	x^6	butterfly
2	4	$x^3 - xy^2$	elliptic umbilic
	4	$x^3 + y^3$	hyperbolic umbilic
	5	$x^2 y + y^4$	parabolic umbilic

The exact meaning of the statement will become clear as we proceed. The Morse Lemma give us a complete classification of germs of corank 0. We shall now classify all germs of corank 1 having finite codimension.

(4.6) Let $f \in \mathcal{M}_n^2$ have corank 1, and finite codimension $(k - 1)$: then

$$f(x_1, \ldots, x_n) \sim \pm x_1^k \pm x_2^2 \pm \ldots \pm x_n^2 .$$

Proof   We know from §3 that a germ of finite codimension is finitely-determined, so the Splitting Lemma applies and tells us that

$$f(x_1, \ldots, x_n) \sim g(x_1) \pm x_2^2 \pm \ldots \pm x_n^2$$

where $g \in \mathcal{M}_1^3$ also has codimension k. Write $x = x_1$. It suffices now to show that $g \sim \pm x^k$. Observe that there must exist a least integer j with

$g \in \mathcal{M}_1^j$, $g \notin \mathcal{M}_1^{j+1}$: if not then $J_g \subseteq \mathcal{M}_1^j$ for all j, so cod $g \geq$ cod $\mathcal{M}_1^j = j$ for all j, i.e. g would have infinite codimension. The Hadamard Lemma tells us that $g(x) = x^j h(x)$ with $h(0) \neq 0$: computation shows that g has codimension $(j - 1)$, so $j = k$. Suppose k odd. In that case take $\phi : (\mathbb{R}, 0) \to (\mathbb{R}, 0)$ to be the germ given by $\phi(x) = x h(x)^{1/k}$: clearly, $\phi'(0) \neq 0$ so ϕ is invertible. Further, $g(x) = \phi^k(x)$, so $g \sim x^k$. The same reasoning applies when k is even and $h(0) > 0$, and yields $g \sim x^k$. But when k is even and $h(0) < 0$ take $\phi(x) = x(-h(x))^{1/k}$ to deduce $g \sim -x^k$. □

It is the germs obtained in the above result which are called the <u>cuspoids</u>. Of course, if we restrict ourselves to germs of codimension ≤ 5 in the result, then we obtain (up to addition of a non-degenerate quadratic form in further variables, and multiplication by ± 1) precisely the germs x^3, x^4, x^5, x^6. And with this we have proved half of Thom's classification theorem (4.5).

A singular point of a smooth function of corank ≤ 1 and finite codimension k is usually called an A_k <u>singularity</u>. Such singularities have long been in the repertoire of classical algebraic geometry, and the reader may find it of interest to recall how they arise naturally when studying double points of algebraic curves.

We shall restrict our attention to real algebraic curves, i.e. subsets of the plane $\mathbb{R}^2$ defined by an equation $f(x, y) = 0$ where $f : \mathbb{R}^2 \to \mathbb{R}$ is a polynomial function: we shall suppose that the curve $f = 0$ passes through the origin (so f has no constant term), and indeed that the origin is a singular point of f (so f has no linear terms either). Recall that the singular point at the origin is said to be of <u>multiplicity</u> m when m is the lowest degree of a monomial in f: thus $m \geq 2$. When $m = 2$ we have a <u>double point</u>, when $m = 3$ a <u>triple point</u>, and so on. The condition on the origin, that it is a double point of the curve $f = 0$, entails that the singular point there has corank ≤ 1. If we assume the singularity to be of finite codimension then it will be of type A_k for some integer $k \geq 1$, so equivalent to the singularity $\pm x^{k+1} \pm y^2$ by (4.6). Multiplication of f by a non-zero scalar does not affect the curve $f = 0$ so we can suppose the normal form is $x^{k+1} \pm y^2$. And since we are not interested in the case when the curve $f = 0$ is a point we consider only the case $x^{k+1} - y^2$. The cases $k = 1, 2, 3, 4, 5$ give rise to the following pictures, which are traditionally dubbed with the titles indicated.

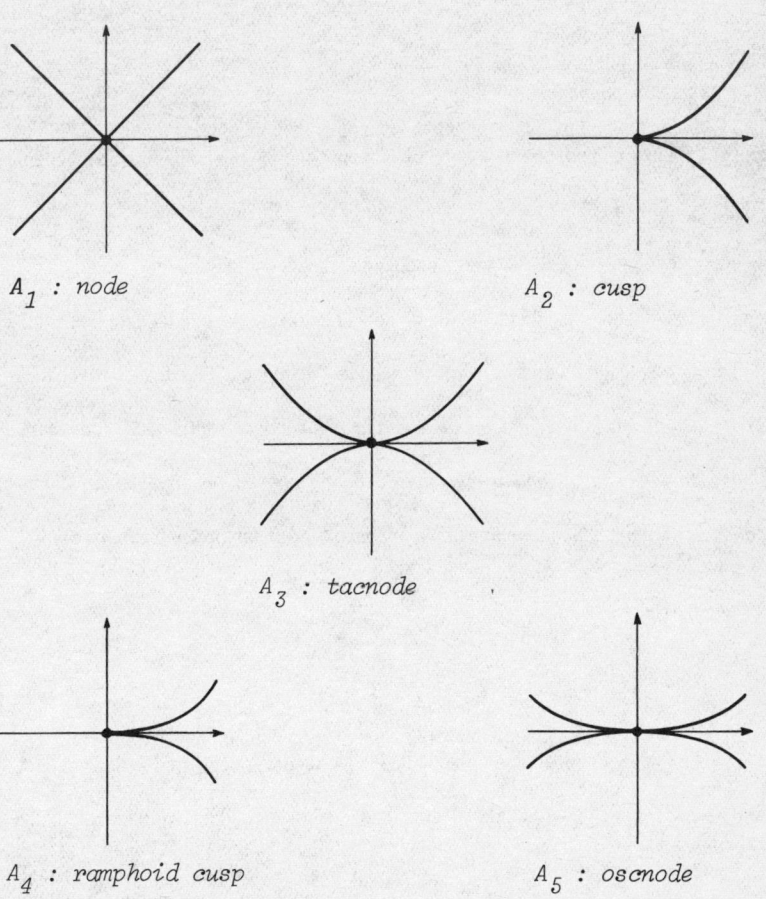

A_1 : node

A_2 : cusp

A_3 : tacnode

A_4 : ramphoid cusp

A_5 : oscnode

<u>Example 1</u> Consider the real algebraic curve given by the equation $f = 0$ where

$$f(x, y) = x^4 + x^2y^2 - 2x^2y - xy^2 + y^2.$$

Clearly, the curve has a singular point at the origin, indeed a double point. And computation verifies that f will have codimension 4 at the origin, so we have a ramphoid cusp. The curve looks like this.

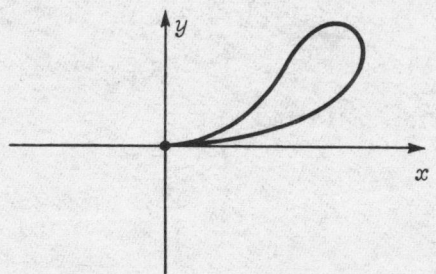

It is perhaps also of interest to see how A_k singularities turn up naturally in studying the differential geometry of plane curves. Suppose we have a smooth mapping $\zeta : \mathbb{R} \to \mathbb{R}^2$ whose image is a smooth submanifold of $\mathbb{R}^2$, with ζ mapping its domain diffeomorphically onto its image: we wish to think of ζ as a parametrization of a plane curve. Let . denote the standard scalar product on $\mathbb{R}^2$, giving rise to the standard distance function $| \ |^2$. (We work with the square to ensure smoothness at the origin.) Given a fixed point w in the plane consider the smooth function $f_w : \mathbb{R} \to \mathbb{R}$ defined by

$$f_w(t) = \tfrac{1}{2}|w - \zeta(t)|$$

measuring, essentially, the distance from w to the point on the curve with parameter t. As w varies we obtain a 2-parameter family $F = (f_w)$ of smooth functions, better thought of as a single smooth function, say $F : \mathbb{R}^2 \times \mathbb{R} \to \mathbb{R}$ given by $F(w, t) = f_w(t)$. Clearly, the singular points of each f_w must have corank ≤ 1, and those of finite codimension will be classified by the A_k-series: indeed the condition for an A_k singularity will be that

$$\frac{\partial F}{\partial t} = 0 \;:\; \frac{\partial^2 F}{\partial t^2} = 0 \;:\; \ldots \;:\; \frac{\partial^k F}{\partial t^k} = 0 \;:\; \frac{\partial^{k+1} F}{\partial t^{k+1}} \neq 0.$$

Let us look at the geometric information contained in these relations. The first few derivatives, written out in full, are

$$\frac{\partial F}{\partial t} = \dot{\zeta}\cdot(\zeta - w)$$

$$\frac{\partial^2 F}{\partial t^2} = \dot{\zeta}\cdot\dot{\zeta} + \ddot{\zeta}\cdot(\zeta - w)$$

$$\frac{\partial^3 F}{\partial t^3} = 3\dot{\zeta}\cdot\ddot{\zeta} + \dddot{\zeta}\cdot(\zeta - w)$$

where $\dot{\zeta}, \ddot{\zeta}, \ldots$ denote the successive derivatives of ζ. Write ζ_1, ζ_2 for the components of ζ. Thus $\dot{\zeta} = (\dot{\zeta}_1, \dot{\zeta}_2)$ is the tangent vector to the curve, and the normal direction is given by $(-\dot{\zeta}_2, \dot{\zeta}_1)$ i.e. the result of rotating the tangent vector through a right angle in an anticlockwise direction. The condition $\frac{\partial F}{\partial t} = 0$ then amounts to saying that the tangent vector $\dot{\zeta}$ is perpendicular to $\zeta - w$, i.e. that w lies on the normal line to the curve at $\zeta(t)$: the set of points (w, t) in $\mathbb{R}^2 \times \mathbb{R}$ where $\frac{\partial F}{\partial t} = 0$ is therefore called the <u>normal bundle space</u> of the curve. Note that for such a pair (w, t) we can write $w = \zeta + \lambda n$ where $n = n(t)$ denotes the unit normal direction to the curve at $\zeta(t)$, and λ is some scalar. Now let us add the further condition that $\frac{\partial^2 F}{\partial t^2} = 0$. Substitute the expression for w just given in that for $\frac{\partial^2 F}{\partial t^2}$, and equate the result to zero; a few lines of computation will then show that

$$\lambda = \frac{\{\dot{\zeta}_1^2 + \dot{\zeta}_2^2\}^{3/2}}{\dot{\zeta}_1\ddot{\zeta}_2 - \ddot{\zeta}_1\dot{\zeta}_2}$$

which the reader will recognize as the radius of curvature, i.e. the reciprocal of the curvature κ. It follows that the points (w, t) in $\mathbb{R}^2 \times \mathbb{R}$ for which $\frac{\partial F}{\partial t} = 0$, $\frac{\partial^2 F}{\partial t^2} = 0$ are precisely those for which $w = \zeta + \frac{1}{\kappa}n$, i.e. with w a point on the evolute of ζ. It is interesting to go one stage further and ask what further geometric information we can glean by adding yet one more condition $\frac{\partial^3 F}{\partial t^3} = 0$: we leave the reader to check that this is equivalent to $\dot{\kappa} = 0$, i.e. to the point $\zeta(t)$ being a vertex of the curve. Thus singularities of type A_k with $k \geq 1$ correspond to the normal bundle space for the curve, those with $k \geq 2$ to points on the evolute, and those with $k \geq 3$ to vertices. It turns out that generically one does not obtain A_k singularities with $k \geq 4$ in this situation, so that is the end of the story.

<u>Example 2</u>   Consider the parabola $\zeta : \mathbb{R} \to \mathbb{R}^2$ given by $\zeta(t) = (t, t^2)$. If we write $w = (u, v)$ then

$$2F(w, t) = t^4 - 2ut - (2v - 1)t^2 + u^2 + v^2$$

and comparison with Example 5 of §1 in Chapter II shows that $\frac{\partial F}{\partial t} = 0$ defines a folded surfaces in $\mathbb{R}^2 \times \mathbb{R}$, that the further condition $\frac{\partial^2 F}{\partial t^2} = 0$ defines the fold curve on the surface, and that the final requirement $\frac{\partial^3 F}{\partial t^3} = 0$ defines the exceptional point in the fold curve where the two folds meet. Under the projection $\mathbb{R}^2 \times \mathbb{R} \xrightarrow{\pi} \mathbb{R}^2$ the fold curve maps to the evolute of the parabola, which is a cuspidal cubic, the cusp arising from the vertex of the parabola at the origin.

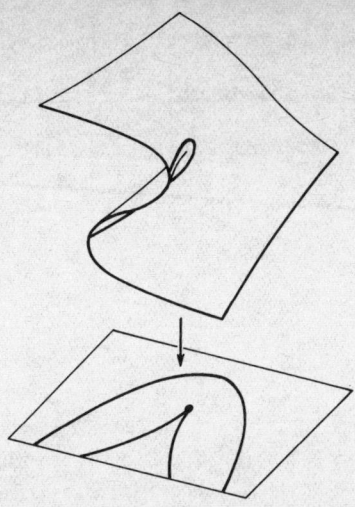

It remains for us to deal with germs of corank 2 and codimension ≤ 5. The next result will complete Thom's classification theorem.

(4.7)   Let $f \in \mathcal{M}_n^2$ have corank 2 and codimension ≤ 5: then f is equivalent to one of the following germs

$$\pm (x_1^3 - x_1 x_2^2) \pm x_3^2 \pm \ldots \pm x_n^2$$
$$\pm (x_1^3 + x_2^3) \pm x_3^2 \pm \ldots \pm x_n^2$$
$$\pm (x_1^2 x_2 + x_2^4) \pm x_3^2 \pm \ldots \pm x_n^2 .$$

Proof   Again, it is the Splitting Lemma which provides us with the first step. It tells us that f is equivalent to a germ

$$g(x_1, x_2) \pm x_3^2 \pm \ldots \pm x_n^2$$

with $g \in \mathcal{M}_2^3$ having the same codimension as f, namely ≤ 5. The third Taylor polynomial of g is a binary cubic form in x_1, x_2 so can be assumed - by the discussion in the previous chapter - to be 0, $x_1^3 - x_1 x_2^2$, $x_1^3 + x_2^3$, $x_1^2 x_2$ according as it is identically zero, symbolic, elliptic, hyperbolic or parabolic. We have to decide which possibilities can arise. To this end, recall that a germ of corank 2 has codimension ≥ 4. In particular f (and hence g) has codimension 4 or 5. We shall consider the possibilities separately.

g has codimension 4

Since $g \in \mathcal{M}^3$ we have $J_g \subseteq \mathcal{M}^2$, so $\mathcal{M} J_g \subseteq \mathcal{M}^3$. In fact $\mathcal{M} J_g = \mathcal{M}^3$. To see this we argue as follows. Put $I = \mathcal{M} J_g$. We are given that J_g has codimension 4, and hence by (2.14) that I has codimension 6. Also, from (2.8) we know that

$$\text{cod } I = \text{cod}_0 I + \text{cod}_1 I + \text{cod}_2 I + \ldots .$$

But $I \subseteq \mathcal{M}^j$, and hence $I + \mathcal{M}^j = \mathcal{M}^j$ for $j \leq 3$, so that

$$\text{cod}_0 I = 1 \ : \ \text{cod}_1 I = 2 \ : \ \text{cod}_2 I = 3$$

we deduce that

$$\text{cod}_3 I = 0$$

which means that $\mathcal{M}^3 \subseteq I = \mathcal{M} J_g$, as required.

It follows now from (3.1) that g is 3-determined so equivalent to its third Taylor polynomial. By checking the five possible normal forms for this one by one we see that the only ones of codimension 4 are the elliptic and hyperbolic types. And these give us the first two germs on the list.

g has codimension 5

Again, since $g \in \mathcal{M}^3$ we have $J_g \subseteq \mathcal{M}^2$ so that $\mathcal{M} J_g \subseteq \mathcal{M}^3$. And again we put $I = \mathcal{M} J_g$. This time J_g has codimension 5 so by (2.14) the codimension of I is 7. Further, we know that

$$\text{cod } I = \text{cod}_0 I + \text{cod}_1 I + \text{cod}_2 I + \text{cod}_3 I + \ldots .$$

As before

$$\text{cod}_0 0 = 1 : \text{cod}_1 I = 2 : \text{cod}_2 I = 3$$

so we must have

$$\text{cod}_3 I = 1 \ \& \ \text{cod}_4 I = 0.$$

By checking the five normal forms for the third Taylor polynomial one by one we see that the only one for which $\text{cod}_3 I = 1$ is the parabolic type $x_1^2 x_2$. And the fact that $\text{cod}_4 I = 0$ tells us that $\mathscr{M}^4 \subseteq \mathscr{M} J_g$ so g is 4-determined by (3.1), hence equivalent to its fourth Taylor polynomial. This has the form

$$x_1^2 x_2 + \Phi(x_1, x_2)$$

where Φ is a homogeneous polynomial of degree 4 in x_1, x_2. In more detail, we can write this as

$$x_1^2 x_2 + \{a x_2^4 + b x_2^3 x_1 + x_2^2 \phi(x_1, x_2)\}$$

where $\phi(x_1, x_2)$ is a quadratic form in x_1, x_2. Notice that if we change co-ordinates, and take the fourth Taylor polynomial of the result, we shall obtain an equivalent germ, since 4-determinacy is invariant under equivalence. In particular, if we replace x_1 by $x_1 - \frac{b}{2} x_2^2$, x_2 by $x_2 - \phi(x_1, x_2)$, and take the fourth Taylor polynomial of the result, we obtain the equivalent germ

$$x_1^2 x_2 + a x_2^4 .$$

Notice that necessarily $a \neq 0$ (for otherwise the germ has infinite codimension). Finally, if we replace x_1 by px_1, x_2 by qx_2, and choose p, q suitably we see that our germ is equivalent to

$$x_1^2 x_2 \pm x_2^4$$

(the sign depending on that of a). And that gives us the last germ on the list.
□

Of course, by combining the results (4.6) and (4.7) we have proved (4.5) and hence obtained the complete classification of germs of codimension ≤ 5.

V Stable singularities of smooth mappings

§1. The Basic Ideas

We come now to the general problem of discussing singular points of smooth mappings. Although the basic geometric ideas are similar to those for functions, the fact is that we are in a much more complex situation and cannot hope to achieve anything like the results of the previous chapter. However, we shall see that it is possible to discuss some decidedly interesting situations if one is prepared to accept one deep result without proof.

The basic ideas are soon set up. Recall that two germs f_1, f_2 are equivalent when there exist invertible germs g, h for which the following diagram commutes.

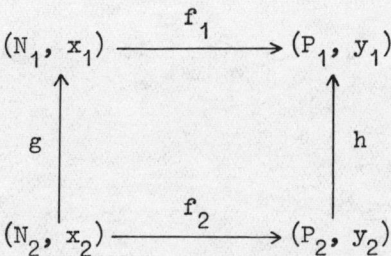

Sometimes it will be convenient to say in this situation that f_1, f_2 are A-equivalent: the reason for this is that later we shall wish to maintain a clear distinction between this notion of equivalence and another to be introduced in §2. Our problem is to classify germs under A-equivalence. We can at least make a start by setting up things in the right way, as we did for functions. The basic objects are germs $f : (N, x) \to (P, y)$ and taking charts at x, y we can suppose $N = \mathbb{R}^n$, $x = 0$ and $P = \mathbb{R}^p$, $y = 0$.

Thus we wish to study the set $\mathcal{E}_{n,p}^0$ of all germs $(\mathbb{R}^n, 0) \to (\mathbb{R}^p, 0)$; note that $\mathcal{E}_{n,p}^0$ has the structure of a real vector space induced from that on $\mathbb{R}^p$. As in Chapter IV we take $\mathcal{R}_n$ to be the set of invertible germs $(\mathbb{R}^n, 0) \to (\mathbb{R}^n, 0)$, which is a group under the operation of composition. Write $A_{n,p} = \mathcal{R}_n \times \mathcal{R}_p$ for the product group. We then have an action $A_{n,p} \times \mathcal{E}_{n,p}^0 \to \mathcal{E}_{n,p}^0$ given by the formula $(g, h).f = h \circ f \circ g^{-1}$, and and the problem of classifying germs in $\mathcal{E}_{n,p}^0$ under this relation of equivalence is precisely the problem of classifying the orbits under this action.

One might hope to proceed just as we did in the previous chapter and attempt to classify germs of low "codimension". But, as we have already said, we are now in a much more complex situation, and it turns out that such a programme is far too ambitious. What we shall do, to make at least some progress, is to restrict ourselves to the germs of lowest possible "codimension", i.e. the "stable" germs in the language of Chapter III. Even this limited objective will prove too much for us, but we shall manage to discuss some very special cases which will suffice to yield complete lists when the dimensions n, p are fairly small.

Our immediate problem is to say just what we mean by a "stable" germ. What we shall do is to proceed by analogy with Chapter III and introduce this notion via that of an "unfolding". The heuristics of the matter proceed as follows. One would like to think of an r-parameter "unfolding" of $f_0 : (\mathbb{R}^m, 0) \to (\mathbb{R}^q, 0)$ as a "germ" $(\mathbb{R}^r, 0) \to (\mathcal{E}_{m,q}^0, f_0)$, say $u \to f_u$. In a natural way that determines $f : (\mathbb{R}^r \times \mathbb{R}^m, 0) \to (\mathbb{R}^q, 0)$ given by $(u, x) \to f_u(x)$, and this in turn determines $F : (\mathbb{R}^r \times \mathbb{R}^m, 0) \to (\mathbb{R}^r \times \mathbb{R}^q, 0)$ given by $(u, x) \to (u, f(u, x))$. On the basis we introduce the following formal definitions. An <u>r-parameter deformation</u> of $f_0 : (\mathbb{R}^m, 0) \to (\mathbb{R}^q, 0)$ is a germ $f : (\mathbb{R}^r \times \mathbb{R}^m, 0) \to (\mathbb{R}^q, 0)$ with $f(0, x) = f_0(x)$ and the corresponding <u>r-parameter unfolding</u> is the germ

$F : (\mathbb{R}^r \times \mathbb{R}^m, 0) \to (\mathbb{R}^r \times \mathbb{R}^q, 0)$ given by $F(u, x) = \left(u, f(u, x)\right)$.

There is precious little difference between the two notions just introduced, and the choice of which one to work with is largely a matter of convenience. In this section we prefer to work with unfoldings, but later in this chapter it will prove more convenient to work with deformations.

<u>Example 1</u>   Let $f : (\mathbb{R}, 0) \to (\mathbb{R}^2, 0)$ be a germ. Think of this as a (small part of a) curve through the origin in the plane. Associated to f we have the distance squared function, $|f|^2$. A 2-parameter deformation of f is the germ $F : (\mathbb{R}^2 \times \mathbb{R}, 0) \to (\mathbb{R}, 0)$ given by

$$F(u, x) = |f(x) - u|^2.$$

As we saw in Chapter IV one can gain interesting geometric information about the curve by studying the singularities of the deformation F.

<u>Example 2</u>   Write $u_1, \ldots, u_{n-1}$ for the standard co-ordinates in $\mathbb{R}^{n-1}$, x for that in $\mathbb{R}$. The germ $F : (\mathbb{R}^{n-1} \times \mathbb{R}, 0) \to (\mathbb{R}^{n-1} \times \mathbb{R}, 0)$ with components $F_1, \ldots, F_n$ where

$$F_i(u, x) = u_i \quad (1 \leq i \leq n - 1)$$

$$F_n(u, x) = x^{n+1} + \sum_{i=1}^{n-1} u_i x^i$$

is an $(n - 1)$-parameter unfolding of $f : (\mathbb{R}, 0) \to (\mathbb{R}, 0)$ given by $f(x) = x^{n+1}$.

Let us now head directly for our initial objective of introducing "stable" germs. For this we need the analogy, in the present situation, of just one of the notions for unfoldings introduced in Chapter III, namely that of "equivalence". Based on the Finite Dimensional Model we make the following definition. Two r-parameter unfoldings F_1, F_2 of a germ $f: (\mathbb{R}^m, 0) \to (\mathbb{R}^q, 0)$ are said to be <u>A-equivalent</u> when there exist r-parameter unfoldings I_m, I_q of the germs at 0 of the identity maps on $\mathbb{R}^m$, $\mathbb{R}^q$ respectively for which the following diagram commutes.

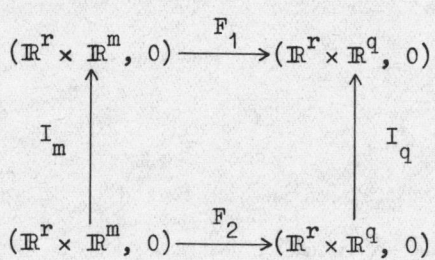

In this situation one refers to the pair $I = (I_m, I_q)$ as an <u>A-equivalence</u> between F_1, F_2. Note that I_m, I_q are necessarily invertible germs. Further, we shall call an r-parameter unfolding F of f <u>A-trivial</u> (or just <u>trivial</u>) when it is A-equivalent to the constant r-parameter unfolding $(\mathbb{R}^r \times \mathbb{R}^m, 0) \to (\mathbb{R}^r \times \mathbb{R}^q, 0)$ of f given by $(u, x) \to (u, f(x))$. And finally we call f <u>A-stable</u> (or just <u>stable</u>) when every unfolding of f is trivial.

We now have a sensible definition of a stable germ; but we have as yet no idea how to produce explicit examples of such germs, and even less of how to go about classifying them. The key to these questions lies in a very simple and beautiful idea whose discussion must of necessity be postponed till §4. The next two sections are devoted to the necessary preliminaries.

§2. Contact Equivalence

The object of this section is to introduce and study another natural equivalence relation on germs $(N, x) \to (P, y)$ with N, P smooth manifolds. This new relation, called $\mathcal{K}$-equivalence, is coarser than $\mathcal{A}$-equivalence and has the advantage that it is fairly computable: in §4 we shall see that there are intimate connexions between the two relations directly relevant to the problem of classifying stable germs.

Suppose then that we have germs $f_1 : (N_1, x_1) \to (P_1, y_1)$ and $f_2 : (N_2, x_2) \to (P_2, y_2)$. We turn our attention to the graphs of representatives, which will be smooth submanifolds of $N_1 \times P_1$, $N_2 \times P_2$ through the points $z_1 = (x_1, y_1)$, $z_2 = (x_2, y_2)$. What we wish to do is to formalize the idea of these graphs having the same "contact" with N_1, N_2 at z_1, z_2 respectively. To this end we define a <u>contact equivalence</u> (or a $\mathcal{K}$-<u>equivalence</u>) to be a pair (h, H) of invertible germs for which one has a commuting diagram

$$\begin{array}{ccccc} (N_1, x_1) & \xrightarrow{i_1} & (N_1 \times P_1, z_1) & \xrightarrow{\Pi_1} & (N_1, x_1) \\ \downarrow h & & \downarrow H & & \downarrow h \\ (N_2, x_2) & \xrightarrow{i_2} & (N_2 \times P_2, z_2) & \xrightarrow{\Pi_2} & (N_2, x_2) \end{array}$$

where for $k = 1, 2$

i_k = germ at x_k of the inclusion $N_k \to N_k \times P_k$
given by $x \mapsto (x, y_k)$

Π_k = germ at z_k of the projection $N_k \times P_k \to N_k$
given by $(x, y) \mapsto x$.

In other words, H is necessarily given by a formula

$$H(x, y) = \bigl(h(x), \theta(x, y)\bigr)$$

with $\theta(x, y_1) = y_2$. And we call our germs f_1, f_2 <u>contact-equivalent</u> (or $\mathcal{K}$-equivalent) when there exists a contact-equivalence (h, H) for which

$$H_0(1, f_1) = (1, f_2) \circ h.$$

Thus one pictures H as mapping the graph of f_1 onto the graph of f_2, with N_1 mapping onto N_2 via h.

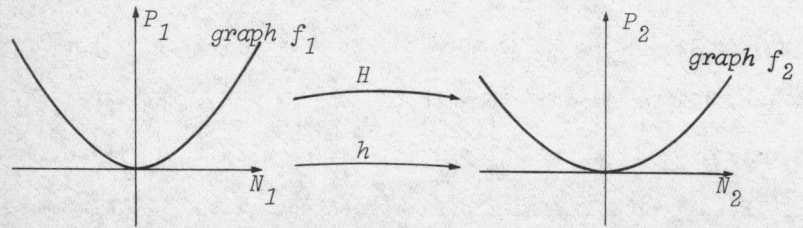

The reader will have no difficulty in checking that if f_1, f_2 are A-equivalent then automatically they are $\mathcal{K}$-equivalent. In particular, when discussing $\mathcal{K}$-equivalence of germs $(N, x) \to (P, y)$ it will be no restriction to suppose that $N = \mathbb{R}^n$, $x = 0$ and $P = \mathbb{R}^p$, $y = 0$. On the other hand as we shall see two germs which are $\mathcal{K}$-equivalent are not necessarily A-equivalent. In the remainder of this section we shall justify our statement that $\mathcal{K}$-equivalence is a fairly computable notion by deriving a more-or-less algebraic criterion for two germs to be $\mathcal{K}$-equivalent. We do this in two steps. The first step, which is the main one, is to look at a finer notion of equivalence called

$\mathcal{C}$-Equivalence.

For the sake of simplicity let us restrict our attention to germs $(\mathbb{R}^n, 0) \to (\mathbb{R}^p, 0)$. Two such germs f, g are <u>$\mathcal{C}$-equivalent</u> when they are $\mathcal{K}$-equivalent via a $\mathcal{K}$-equivalence (h, H) having the special property that h

is the germ at 0 of the identity mapping on $\mathbb{R}^n$. Note therefore that f, g are $\mathcal{K}$-equivalent if and only if there exists an invertible germ $h : (\mathbb{R}^n, 0) \to (\mathbb{R}^n, 0)$ for which $f \circ h$, g are $\mathcal{C}$-equivalent. The next proposition provides us with a purely algebraic criterion for f, g to be $\mathcal{C}$-equivalent. To simplify notation write $I_f = f^*(\mathcal{M}_p)$ i.e. the ideal in $\mathcal{E}_n$ generated by the components $f_1, \ldots, f_p$ of f. And similarly for I_g.

(2.1) <u>The following three conditions on germs f, g in $\mathcal{E}_{n,p}^0$ are equivalent</u>.

    (i)   <u>f, g are $\mathcal{C}$-equivalent</u>.

    (ii) <u>The ideals I_f, I_g are equal</u>.

    (iii) <u>There exists an invertible $p \times p$ matrix (u_{ij}) with entries in $\mathcal{E}_n$ for which $f_i = \sum_j u_{ij} g_j$ for $1 \leq i \leq p$</u>.

It will be convenient to split the proof into a number of fairly easy steps.

Step 1   In which we show that (i) implies (ii). Suppose f, g are $\mathcal{C}$-equivalent. It will suffice to establish $I_f \subseteq I_g$, since then $I_g \subseteq I_f$ follows by symmetry. We must therefore show that each component f_i can be written in the form

$$f_i = \sum_j a_{ij} g_j \qquad (1)$$

with the a_{ij} in $\mathcal{E}_n$. Since f, g are $\mathcal{C}$-equivalent there exists a $\mathcal{C}$-equivalence $(1, H)$ for which $H(x, g(x)) = (x, f(x))$. Write $H(x, y) = (x, \theta(x, y))$ where θ satisfies $\theta(x, 0) = 0$ identically in x, and hence its components $\theta_1, \ldots, \theta_p$ satisfy $\theta_i(x, 0) = 0$ identically in x. By the Hadamard Lemma we can write

$$\theta_i(x, y) = \sum_j y_j \theta_{ij}(x, y)$$

with $y_1, \ldots, y_p$ the standard co-ordinates in $\mathbb{R}^p$, and the θ_{ij} in $\mathcal{E}_n$. But now

$$f_i(x) = \theta_i(x, g(x)) = \sum_j g_j(x) \theta_{ij}(x, g(x))$$

which has the desired form (1) if we put $a_{ij}(x) = \theta_{ij}(x, g(x))$.

Step 2 In which we isolate a lemma from linear algebra to be used in the next step. We claim that given real $p \times p$ matrices A, B there exists a real $p \times p$ matrix C for which $U = C(I - AB) + B$ is invertible. (Here I is the identity $p \times p$ matrix.) To this end let $a, b : \mathbb{R}^p \to \mathbb{R}^p$ be the linear mappings whose matrices, relative to the standard bases, are respectively A, B. Choose a basis $e_1, \ldots, e_p$ for $\mathbb{R}^p$ for which $b(e_i) = 0$ for $r + 1 \leq i \leq p$, where r denotes the rank of B: then $b(e_1), \ldots, b(e_r)$ are linearly independent vectors in $\mathbb{R}^p$, so can be extended by vectors $e'_{r+1}, \ldots, e'_p$ to a basis for $\mathbb{R}^p$. Now let $c : \mathbb{R}^p \to \mathbb{R}^p$ be the linear mapping defined by $c(e_i) = 0$ for $1 \leq i \leq r$, and $c(e_i) = e'_i$ for $r + 1 \leq i \leq p$. Observe that $u = c(1 - ab) + b$ has the property that for some scalars λ_{ij} one has

$$u(e_i) = \begin{cases} b(e_i) + \sum_{j=r+1}^{p} \lambda_{ij} e'_i & \text{for } 1 \leq i \leq r \\ e'_i & \text{for } r + 1 \leq r \leq p \end{cases}$$

so u is invertible, and the matrix C of c is the required matrix.

Step 3   In which we show that (ii) implies (iii). Assume that the ideals I_f, I_g are equal, so we can certainly write

$$g_i = \sum_j a_{ij} f_j$$

$$f_i = \sum_j b_{ij} g_j$$

where the coefficients a_{ij}, b_{ij} lie in $\mathcal{E}_n$. For each x let A_x, B_x be the real $p \times p$ matrices with entries $a_{ij}(x)$, $b_{ij}(x)$ respectively. By Step 2 we can find a real $p \times p$ matrix C_0 with $U_0 = C_0(I - A_0 B_0) + B_0$ invertible. It follows that for x close to 0 the matrix $U_x = C_0(I - A_x B_x) + B_x$ is invertible as well. Let $u_{ij}(x)$ be the entries of U_x. It is now a matter of computation to check that

$$f_i = \sum_j u_{ij} g_j .$$

Step 4   In which we complete the proof of (2.1) by showing that (iii) implies (i). Suppose there exists an invertible $p \times p$ matrix (u_{ij}) with the properties stated in (iii). Let us define a germ $\theta : (\mathbb{R}^n \times \mathbb{R}^p, 0) \to (\mathbb{R}^p, 0)$ by taking its components to be

$$\theta_i(x, y) = \sum_j y_j u_{ij}(x) \quad (1 \leq i \leq p).$$

Certainly then $\theta(x, 0) = 0$ identically in x. Now define a germ $H : (\mathbb{R}^n \times \mathbb{R}^p, 0) \to (\mathbb{R}^n \times \mathbb{R}^p, 0)$ by the formula $H(x, y) = \bigl(x, \theta(x, y)\bigr)$. One checks easily that the Jacobian matrix of H at 0 is invertible, so H is an invertible germ, and $(1, H)$ is a $\mathcal{C}$-equivalence. Moreover, by construction $\theta_i\bigl(x, g(x)\bigr) = f_i(x)$ so $\theta\bigl(x, g(x)\bigr) = f(x)$ and $H\bigl(x, g(x)\bigr) = \bigl(x, f(x)\bigr)$ entailing that f, g are $\mathcal{C}$-equivalent. □

Example 1   The germs $f(x) = (x^2, 0)$ and $g(x) = (x^2, x^3)$ are $\mathcal{C}$-equivalent, since the ideals generated by the components clearly coincide, and hence $\mathcal{K}$-equivalent. On the other hand f, g are not $\mathcal{A}$-equivalent. Suppose indeed that we could find invertible germs $h : (\mathbb{R}, 0) \to (\mathbb{R}, 0)$ and $k : (\mathbb{R}^2, 0) \to (\mathbb{R}^2, 0)$. for which $f \circ h = k \circ g$. Working with representatives, we see that k would map $\text{Im} f - \{0\}$ to $\text{Im} g - \{0\}$: that however is impossible since the former set is connected, whilst the latter is disconnected.

Example 2   Consider a germ $f : (\mathbb{R}, 0) \to (\mathbb{R}, 0)$ for which the following conditions are satisfied.

$$\frac{\partial f}{\partial x}(0) = 0 : \ldots : \frac{\partial^k f}{\partial x^k}(0) = 0 : \frac{\partial^{k+1} f}{\partial x^{k+1}}(0) \neq 0.$$

It follows that $f = x^{k+1} g$ for some $g : (\mathbb{R}, 0) \to (\mathbb{R}, 0)$ with $g(0) \neq 0$. (The argument appeared twice in Chapter IV.) Now g is invertible in the algebra $\mathcal{E}_1$ so $\langle f \rangle = \langle x^{k+1} \rangle$ and we conclude that f is $\mathcal{C}$-equivalent to x^{k+1}.

Example 3   Let $f, g : (\mathbb{R}^n, 0) \to (\mathbb{R}^p, 0)$ be germs of linear mappings, with components f_i, g_i respectively. If these linear mappings have the same image we can write each f_i as a linear combination of $g_1, \ldots, g_p$ with scalar coefficients, and similarly each g_i as a linear combination of $f_1, \ldots, f_p$ with scalar coefficients: certainly then the ideals I_f, I_g coincide, so f, g are $\mathcal{C}$-equivalent. Conversely, we claim that if f, g are $\mathcal{C}$-equivalent then they have the same image. Indeed in this case we can write each f_i as a linear combination of $g_1, \ldots, g_p$ with coefficients in $\mathcal{E}_n$, and each g_i as a linear combination of $f_1, \ldots, f_p$ with coefficients in $\mathcal{E}_n$. However as the f_i, g_i are linear, the coefficients must be scalars so f, g give rise to the same image.

Back to $\mathcal{K}$-Equivalence

Although we managed to produce a purely algebraic criterion for two germs to be $\mathcal{C}$-equivalent, we have to be content with somewhat less for $\mathcal{K}$-equivalence. As a preliminary, call two ideals in $\mathcal{E}_n$ <u>isomorphic</u> when there exists an algebra isomorphism of $\mathcal{E}_n$ to itself which maps the one ideal onto the other. We need a more special notion. Two ideals in $\mathcal{E}_n$ are <u>induced isomorphic</u> when there exists an isomorphism of $\mathcal{E}_n$ to itself, of the form h^* for some germ $h : (\mathbb{R}^n, 0) \to (\mathbb{R}^n, 0)$, which maps the one ideal onto the other. Note that by (IV.2.11) the germ h is necessarily invertible. We contend that

(2.2)   <u>A necessary and sufficient condition for two germs</u> $f, g : (\mathbb{R}^n, 0) \to (\mathbb{R}^p, 0)$ <u>to be $\mathcal{K}$-equivalent is that the ideals</u> I_f, I_g <u>are induced isomorphic</u>.

<u>Necessity</u>   Suppose f, g are $\mathcal{K}$-equivalent, so there exists an invertible germ $h : (\mathbb{R}^n, 0) \to (\mathbb{R}^n, 0)$ for which $f \circ h, g$ are $\mathcal{C}$-equivalent. It follows from (2.1) that the ideals $I_{f \circ h}$, I_g are equal. The induced isomorphism h^* will map I_f onto I_g, so these ideals are induced isomorphic.

<u>Sufficiency</u>   Suppose I_f, I_g induced isomorphic, so there exists an induced isomorphism h^* of $\mathcal{E}_n$, with $h : (\mathbb{R}^n, 0) \to (\mathbb{R}^n, 0)$ an invertible germ, which maps I_f onto I_g. That means the ideals $I_{f \circ h}$, I_g are equal, so by (2.1) the germs $f \circ h, g$ are $\mathcal{C}$-equivalent, and hence f, g are $\mathcal{K}$-equivalent.
□

<u>Example 4</u>   We have already observed that the germs $f(x, y) = (x^2, y^2)$ and $g(x, y) = (x^2 + y^2, xy)$ are A-equivalent hence $\mathcal{K}$-equivalent. On the

other hand these germs are not $\mathcal{C}$-equivalent since (for instance) the component xy does not lie in the ideal generated by x^2, y^2.

Thus far we have vindicated our claim that $\mathcal{K}$-equivalence is a fairly computable notion. And being a coarser relation than A-equivalence one has rather more hope of being able to classify germs under $\mathcal{K}$-equivalence. In §4 we shall see that for stable germs the two notions coincide, and this is the reason why we are going to pursue the problem of classifying germs under $\mathcal{K}$-equivalence.

The first step in this programme is to mimic the approach adopted in Chapter IV to deal with germs of functions, i.e. we shall set up the problem as one of classifying the orbits under a group action, and then proceed by analogy with the Finite Dimensional Model.

This goes as follows. Observe first that the set $\mathcal{K}_{n,p}$ of all $\mathcal{K}$-equivalences (h, H) with $h : (\mathbb{R}^n, 0) \to (\mathbb{R}^n, 0)$ and
$H : (\mathbb{R}^n \times \mathbb{R}^p, 0) \to (\mathbb{R}^n \times \mathbb{R}^p, 0)$ forms a group, the operation being given by composition of the H. Note that the set $\mathcal{C}_{n,p}$ of all $\mathcal{C}$-equivalences $(1, H)$ is a subgroup of $\mathcal{K}_{n,p}$: also that we can identify the group $\mathcal{R}_n$ discussed in Chapter III with the subgroup of $\mathcal{K}_{n,p}$ comprising all $\mathcal{K}$-equivalences $(h, h \times 1)$. With this identification it is clear that any element (h, H) in $\mathcal{K}_{n,p}$ can be written uniquely as the product of the element $(h, h \times 1)$ in $\mathcal{R}_n$ and the element $(1, H')$ in $\mathcal{C}_{n,p}$ where $H' = (h \times 1)^{-1} \circ H$: one expresses this by saying that $\mathcal{K}_{n,p}$ is the <u>semi-direct product</u> of $\mathcal{R}_n$ and $\mathcal{C}_{n,p}$. The group $\mathcal{K}_{n,p}$ acts on $\mathcal{C}_{n,p}^0$ by agreeing that

$$\left(1, (h, H).f\right) = H \circ (1, f) \circ h^{-1}$$

and it should be clear that two germs $(\mathbb{R}^n, 0) \to (\mathbb{R}^p, 0)$ lie in the same orbit under this action if and only if they are $\mathscr{K}$-equivalent. Thus the problem of classifying germs under $\mathscr{K}$-equivalence is that of classifying the orbits under the above action.

Now we proceed as in Chapter IV. We pretend that the action of $\mathscr{K}_{n,p}$ on $\mathscr{E}^0_{n,p}$ is that of a Lie group on a smooth manifold, and look for a candidate for the "tangent space" to the orbit $\mathscr{K} \cdot f$ through the germ $f : (\mathbb{R}^n, 0) \to (\mathbb{R}^p, 0)$. We expect this to be the image of the "differential" at the identity of the natural mapping of the group onto the orbit through f, which we can think of as a mapping $\mathscr{R}_n \times \mathscr{C}_{n,p} \to \mathscr{E}^0_{n,p}$. Since the domain is a product we expect this image to be the vector sum of the images of the "differentials" at the identities of the component mappings $\mathscr{R}_n \to \mathscr{E}^0_{n,p}$ and $\mathscr{C}_{n,p} \to \mathscr{E}^0_{n,p}$. We consider these separately.

The Mapping $\mathscr{R}_n \to \mathscr{E}^0_{n,p}$

This is given by the formula $h \to f \circ h^{-1}$. One reasons exactly as in Chapter IV. Write $f_1, \ldots, f_p$ for the components of f, and $\frac{\partial f}{\partial x_i}$ for the germ with components $\frac{\partial f_1}{\partial x_i}, \ldots, \frac{\partial f_p}{\partial x_i}$ where $x_1, \ldots, x_n$ denote the standard co-ordinates in $\mathbb{R}^n$. We expect the required "differential" at the identity to be the linear mapping

$$(g_1, \ldots, g_n) \to g_1 \frac{\partial f}{\partial x_1} + \ldots + g_n \frac{\partial f}{\partial x_n} \qquad *$$

with each g_i a germ $(\mathbb{R}^n, 0) \to (\mathbb{R}, 0)$. For the reasons explained in Chapter IV we do not wish to restrict these germs to have zero target, so allow them to be arbitrary germs in $\mathscr{E}_n$. Thus the image of the linear mapping given by * will be the submodule $\mathscr{E}_n \left\{ \frac{\partial f}{\partial x_1}, \ldots, \frac{\partial f}{\partial x_n} \right\}$ of the $\mathscr{E}_n$-module $\mathscr{E}_{n,p}$: this submodule is called the <u>Jacobian module</u> of f, and is written J_f. Of course in the case $p = 1$ we recover the Jacobian ideal introduced in Chapter IV.

The Mapping $\mathcal{E}_{n,p} \to \mathcal{E}^0_{n,p}$

To simplify life identify a $\mathcal{C}$-equivalence $(1, H)$ where $H(x, y) = \bigl(x, \theta(x, y)\bigr)$ with θ. Then our mapping is given by composition on the right with $(1, f)$ so is the restriction of a linear mapping, and should be its own "differential" at any point. We need therefore the image of this mapping taken over germs $\theta(x, y)$ which vanish identically in x when $y = 0$. Write $\theta_1, \ldots, \theta_p$ for the components of θ. Each component $\theta_i(x, y)$ likewise vanishes identically in x when $y = 0$, so by the Hadamard Lemma must lie in the ideal generated by the (germs at 0 of the) standard co-ordinates $y_1, \ldots, y_p$ on $\mathbb{R}^p$. Thus $\theta_i\bigl(x, f(x)\bigr)$ is a linear combination of $f_1, \ldots, f_p$ with coefficients in $\mathcal{E}_n$, i.e. an element of the ideal I_f generated by the components of f: reversing the steps one sees moreover that every element of I_f can be so obtained. Thus the image of our mapping comprises all germs in $\mathcal{E}_{n,p}$ with components in I_f, i.e. the $\mathcal{E}_n$-submodule $I_f \cdot \mathcal{E}_{n,p}$.

On this heuristic basis we arrive at the following formal definitions. Given a germ $f : (\mathbb{R}^n, 0) \to (\mathbb{R}^p, 0)$ we define the $\mathcal{K}$-tangent space to f to be the $\mathcal{E}_n$-submodule $T_f = J_f + I_f \cdot \mathcal{E}_{n,p}$. And we define the $\mathcal{K}$-codimension of f to be the codimension of this vector subspace in $\mathcal{E}_{n,p}$. In view of (IV.2.7) we have the following criterion for a germ to be of finite $\mathcal{K}$-codimension.

(2.3) <u>A necessary and sufficient condition for a germ</u> $f : (\mathbb{R}^n, 0) \to (\mathbb{R}^p, 0)$ <u>to be of finite $\mathcal{K}$-codimension is that there exists an integer</u> $k \geq 1$ <u>with</u> $\mathcal{M}_n^k \cdot \mathcal{E}_{n,p} \subseteq T_f$.

This is fairly straightforward to apply in practice. Write the components of f as $f_1, \ldots, f_p$ and think of $\mathcal{E}_{n,p}$ as $\mathcal{E}_n \times \ldots \times \mathcal{E}_n$ (p times). Then $I_f \cdot \mathcal{E}_{n,p}$ will be generated by all p-tuples $(0, \ldots, f_i, \ldots, 0)$, with the f_i in any position, whilst J_f is generated by the
$$\frac{\partial f}{\partial x_i} = \left(\frac{\partial f_1}{\partial x_i}, \ldots, \frac{\partial f_p}{\partial x_i}\right):$$
thus T_f is generated by the list of all these vectors. Now $\mathcal{M}_n^k$ is generated by all monomials m of degree k in $x_1, \ldots, x_n$, and hence $\mathcal{M}_n^k \cdot \mathcal{E}_{n,p}$ is generated by all p-tuples $(0, \ldots, m, \ldots, 0)$. And to check the condition $\mathcal{M}_n^k \cdot \mathcal{E}_{n,p} \subseteq T_f$, one has only to check that each such p-tuple can be written as a linear combination (with coefficients in $\mathcal{E}_n$) of the generators for T_f. Doing this successively for $k = 1, 2, 3, \ldots$ there is at least a sporting chance that in a given example one will either find a k for which the condition is satisfied, or see that it cannot be satisfied for any k.

<u>Example 5</u> We shall show that the germ $f(x, y) = (x^2, y^2)$, the germ at 0 of the "folded handkerchief" mapping, has finite $\mathcal{K}$-codimension. Here $\frac{\partial f}{\partial x} = (2x, 0)$, $\frac{\partial f}{\partial y} = (0, 2y)$. Also $I_f \cdot \mathcal{E}_{2,2}$ is generated by $(x^2, 0)$, $(y^2, 0)$, $(0, x^2)$, $(0, y^2)$. Clearly then the $\mathcal{K}$-tangent space T_f is the $\mathcal{E}_2$-submodule of $\mathcal{E}_{2,2}$ generated by $(x, 0)$, $(y^2, 0)$, $(0, y)$ and $(0, x^2)$.

We start by trying to verify the condition of (2.3) in the case $k = 1$. The ideal $\mathcal{M}_i$ is generated by the monomials x, y so the $\mathcal{E}_2$-submodule $\mathcal{M}_2 \cdot \mathcal{E}_{2,2}$ is generated by $(x, 0)$, $(y, 0)$, $(0, x)$, $(0, y)$. The question now is whether each of these four vectors can be written as linear combinations of the four generators we obtained for T_f. For $(x, 0), (0, y)$ this is trivial. But clearly we cannot express $(y, 0), (0, x)$ as linear combinations of the four generators, so the condition fails.

We continue therefore by trying to verify the condition of (2.3) in the case $k = 2$. The ideal $\mathcal{M}_2^2$ is generated by the monomials x^2, xy, y^2 so the $\mathcal{E}_2$-submodule $\mathcal{M}_2^2 \cdot \mathcal{E}_{2,2}$ is generated by $(x^2, 0)$, $(xy, 0)$, $(y^2, 0)$, $(0, x^2)$, $(0, xy)$, $(0, y^2)$. Again, the question is whether each of these vectors can be written as a linear combination of the four generators for T_f. Clearly, this is the case, so $\mathcal{M}_2^2 \cdot \mathcal{E}_{2,2} \subseteq T_f$, and f has finite $\mathcal{K}$-codimension.

Before going any further we should check that the $\mathcal{K}$-codimension of a germ is actually a $\mathcal{K}$-invariant, i.e. that

(2.4)   If two germs $(\mathbb{R}^n, 0) \to (\mathbb{R}^p, 0)$ are $\mathcal{K}$-equivalent then they have the same $\mathcal{K}$-codimension.

Proof   For the purposes of the proof we shall identify $\mathcal{E} = \mathcal{E}_{n,p}$ with the product $\mathcal{E}_n \times \ldots \times \mathcal{E}_n$ (p times) by identifying a germ f in $\mathcal{E}$ with the p-tuple $(f_1, \ldots, f_p)$ of its components relative to the standard coordinates on $\mathbb{R}^n$, $\mathbb{R}^p$. We proceed in steps.

Step 1   Let $u = (u_{ij})$ be an invertible $p \times p$ matrix with entries in $\mathcal{E}_n$. We take $U : \mathcal{E} \to \mathcal{E}$ to be the isomorphism of real vector spaces given by $f \to u \cdot f$, where we think of f as a column vector. Put $g = u \cdot f$. We claim that U maps T_f isomorphically onto T_g. It will suffice to show that $T_g \subseteq U(T_f)$, for then similar reasoning establishes $T_f \subseteq U^{-1}(T_g)$ and hence $U(T_f) \subseteq T_g$. Clearly then the problem reduces to that of establishing the two inclusions. $I_g \cdot \mathcal{E} \subseteq U(T_f)$ and $J_g \subseteq U(T_f)$. For the first, observe that $I_g \cdot \mathcal{E} \subseteq I_f \cdot \mathcal{E} = U(I_f \cdot \mathcal{E}) \subseteq U(T_f)$, the equality following from Cramer's Rule. And for the second observe that
$J_g \subseteq U(J_f) + I_f \cdot \mathcal{E} = U(J_f) + U(I_f \cdot \mathcal{E}) = U(T_f)$, the inclusion following from the rule for differentiating a product of two functions.

Step 2   Let $\phi : (\mathbb{R}^n, 0) \to (\mathbb{R}^n, 0)$ be an invertible germ, and let $\Phi : \mathcal{E} \to \mathcal{E}$ be the isomorphism of real vector spaces given by $f \to f \circ \phi$.

Put $g = f \circ \phi$. We claim that Φ maps T_f isomorphically onto T_g. As in Step 1 it suffices to establish an inclusion $T_g \subseteq \Phi(T_f)$ which reduces to establishing the two inclusions $I_g \cdot \mathcal{E} \subseteq \Phi(T_f)$ and $J_g \subseteq \Phi(T_f)$. The first inclusion follows immediately from the fact that $I_g \cdot \mathcal{E} \subseteq \Phi(I_f \cdot \mathcal{E})$. And the second inclusion follows from the fact that, by the Chain Rule, we have, for $1 \le i \le n$,

$$\frac{\partial g}{\partial x_i} = \sum_{j=1}^{n} \frac{\partial \phi_j}{\partial x_i} \left(\frac{\partial f}{\partial x_j} \circ \phi \right).$$

<u>Step 3</u>   Let $f, h : (\mathbb{R}^n, 0) \to (\mathbb{R}^p, 0)$ be $\mathcal{K}$-equivalent germs. Then there exists an invertible germ ϕ for which $g = f \circ \phi$, h are $\mathcal{E}$-equivalent. And by (2.1) there exists an invertible $p \times p$ matrix $u = (u_{ij})$ with entries in $\mathcal{E}_n$ for which $h = u \cdot g$. It follows from the previous steps that the automorphism $U \circ \Phi$ of the real vector space $\mathcal{E}$ maps T_f isomorphically onto T_h, so the quotient spaces $\mathcal{E}/T_f$, $\mathcal{E}/T_h$ are isomorphic and have the same dimension, i.e. f, h have the same $\mathcal{K}$-codimension.   □

We come now to the question of actually computing the $\mathcal{K}$-codimension of a germ. Before describing a fairly systematic method of doing this it may be worthwhile looking at the familiar case $p = 1$ of germs of functions to see how this differs from the situation studied in Chapter IV. Consider then a germ $f : (\mathbb{R}^n, 0) \to (\mathbb{R}, 0)$. The $\mathcal{K}$-tangent space to f is $J_f + I_f$. A special case arises when $f \in J_f$, since then the $\mathcal{K}$-tangent space reduces to just J_f, and the $\mathcal{K}$-codimension will coincide with the $\mathcal{R}$-codimension as defined in Chapter IV. For instance, this applies to the A_k-singularities $\pm x_1^2 \pm \ldots \pm x_{n-1}^2 \pm x_n^{k+1}$, to the elliptic umbilic $x^3 - xy^2$, to the hyperbolic umbilic $x^3 + y^3$, and to the parabolic umbilic $x^2 y + y^4$. But in general the relation between the two codimensions is not well understood at the time of writing. It is known that a germ of a function is of finite $\mathscr{K}$-codimension if and only if it is of finite $\mathscr{R}$-codimension, though there seems to be no easy proof of this fact.

As in the case of germs of functions, the method we use to compute codimensions of germs of mappings is based upon (IV.2.8). Consider a germ $f : (\mathbb{R}^n, 0) \to (\mathbb{R}^p, 0)$. One writes cod f for the $\mathscr{K}$-codimension of f, or just cod f, when it is quite clear that we are dealing with $\mathscr{K}$-equivalence. The proposition just mentioned tells us that

$$\text{cod } f = \text{cod}_0 f + \text{cod}_1 f + \text{cod}_2 f + \ldots$$

where

$$\text{cod}_k f = \dim \frac{T_f + \mathscr{M}_n^k \cdot \mathscr{E}_{n,p}}{T_f + \mathscr{M}_n^{k+1} \cdot \mathscr{E}_{n,p}} .$$

What we do is to compute the integers $\text{cod}_0 f$, $\text{cod}_1 f$, ... successively, as as follows. Suppose we wish to compute $\text{cod}_k f$. As we have already pointed out we have an explicit finite list of generators for $\mathscr{M}_n^k \cdot \mathscr{E}_{n,p}$ namely the p-tuples $(0, \ldots, m, 0, \ldots, 0)$ with m a monomial of degree k in $x_1, \ldots, x_n$. The first thing one has to do is to check which of these generators lies in $T_f + \mathscr{M}_n^{k+1} \cdot \mathscr{E}_{n,p}$: this usually amounts to doing a small computation for each generator in turn. There is a practical point to note here namely that we need only consider the generators for T_f modulo $\mathscr{M}_n^{k+1} \cdot \mathscr{E}_{n,p}$, which means in practice that for each such generator we can put equal to zero all terms of degree $\geq k + 1$. The next thing to do is to select from the generators which do not lie in $T_f + \mathscr{M}_n^{k+1} \cdot \mathscr{E}_{n,p}$ a basis for a supplement in $T_f + \mathscr{M}_n^k \cdot \mathscr{E}_{n,p}$: the number of basis elements is the number $\text{cod}_k f$. Of course, the process comes to an end when one finds an integer $k \geq 1$ for which all the generators for $\mathscr{M}_n^k \cdot \mathscr{E}_{n,p}$ lie in $T_f + \mathscr{M}_n^{k+1} \cdot \mathscr{E}_{n,p}$. Finally, one adds up the list of integers $\text{cod}_0 f$, $\text{cod}_1 f$, ... to get cod f. Note that if none of the components $f_1, \ldots, f_p$ of f involves linear terms then $T_f \subseteq \mathscr{M}_n \cdot \mathscr{E}_{n,p}$ and hence $\mathrm{cod}_0 f = p$: this remark will apply to all the examples below.

Example 6   We shall compute the $\mathscr{K}$-codimension of the germ $f : (\mathbb{R}^2, 0) \to (\mathbb{R}^2, 0)$ of the "folded handkerchief" mapping given by $f(x, y) = (x^2, y^2)$. As we saw in Example 5 the $\mathscr{K}$-tangent space T_f is the submodule of $\mathscr{E}_{2,2}$ generated by $(x, 0)$, $(y^2, 0)$, $(0, y)$, $(0, x^2)$. Here $\mathrm{cod}_0 f = 2$. To compute $\mathrm{cod}_1 f$ we have first to determine which generators for $\mathscr{M}_2 \cdot \mathscr{E}_{2,2}$ lie in $T_f + \mathscr{M}_2^2 \cdot \mathscr{E}_{2,2}$. Now $\mathscr{M}_2 \cdot \mathscr{E}_{2,2}$ is generated by $(x, 0)$, $(y, 0)$, $(0, x)$, $(0, y)$: of these $(x, 0)$, $(0, y)$ obviously lie in $T_f + \mathscr{M}_2^2 \cdot \mathscr{E}_{2,2}$ whilst $(y, 0)$, $(0, x)$ do not. Indeed these vectors form a supplement for $T_f + \mathscr{M}_2^2 \cdot \mathscr{E}_{2,2}$ in $T_f + \mathscr{M}_2 \cdot \mathscr{E}_{2,2}$ so $\mathrm{cod}_1 f = 2$. We need go no further now because, as we saw in Example 5, we have $\mathscr{M}_2^2 \cdot \mathscr{E}_{2,2} \subseteq T_f$. Thus $\mathrm{cod}_f = 2 + 2 = 4$.

Example 7   Consider the germ $f : (\mathbb{R}, 0) \to (\mathbb{R}^p, 0)$ defined by $f(x) = (0, \ldots, 0, x^{t+1})$ with $t \geq 1$ an integer. Here the Jacobian module J_f is generated by $(0, \ldots, 0, x^t)$, and the ideal I_f is generated by the vectors $(0, \ldots, x^{t+1}, \ldots, 0)$: it follows that the $\mathscr{K}$-tangent space T_f is generated by all vectors $(0, \ldots, x^{t+1}, \ldots, 0)$ where the power does not appear in the last component, and $(0, 0, \ldots, x^t)$. Let us compute $\mathrm{cod}_k f$. The ideal $\mathscr{M}_1^k$ is generated by x^j, so the submodule $\mathscr{M}_1^k \cdot \mathscr{E}_{1,p}$ is generated by the vectors $(0, \ldots, x^k, \ldots, 0)$. We ask which of these generators lies in the submodule $T_f + \mathscr{M}_1^{k+1} \cdot \mathscr{E}_{1,p}$. There are three cases to consider.

The Case $k < t$ None of the generators for $\mathscr{M}_1^k \cdot \mathscr{E}_{1,p}$ lies in $T_f + \mathscr{M}_1^{k+1} \cdot \mathscr{E}_{1,p}$: there are p such vectors, and they form a supplement, so

$\text{cod}_k f = p$ for $k < t$.

The Case $k = t$ None of the generators for $\mathcal{M}_1^k \cdot \mathcal{E}_{1,p}$ lie in $T_f + \mathcal{M}_1^{k+1} \cdot \mathcal{E}_{1,p}$ save the last one $(0, \ldots, x^t)$: there are $(p-1)$ such generators, forming a supplement, so $\text{cod}_k f = p - 1$ when $k = t$.

The Case $k > t$ All the generators for $\mathcal{M}_1^k \cdot \mathcal{E}_{1,p}$ lie in $T_f + \mathcal{M}_1^{k+1} \cdot \mathcal{E}_{1,p}$, so $\text{cod}_k f = 0$ for $k > t$.

We conclude that the germ f has $\mathcal{K}$-codimension $pt + p - 1$.

Example 8   We shall compute the $\mathcal{K}$-codimension of the germ $f : (\mathbb{R}^2, 0) \to (\mathbb{R}^2, 0)$ given by $f(x, y) = (xy, x^a + y^b)$ where a, b are integers ≥ 3. Here $\frac{\partial f}{\partial x} = (y, ax^{a-1})$ and $\frac{\partial f}{\partial y} = (x, by^{b-1})$ generate the Jacobian module J_f. The ideal I_f is generated by xy, $x^a + y^b$ so the submodule $I_f \cdot \mathcal{E}_{2,2}$ is generated by $(xy, 0)$, $(x^a + y^b, 0)$, $(0, xy)$ and $(0, x^a + y^b)$. The reader will find it a straightforward exercise to verify that the $\mathcal{K}$-tangent space is generated by $\frac{\partial f}{\partial x}$, $\frac{\partial f}{\partial y}$ together with $(x^2, 0)$, $(xy, 0)$, $(y^2, 0)$, $(0, x^a)$, $(0, xy)$, $(0, y^b)$. We wish to compute $\text{cod}_k f$ for $k \geq 1$. The ideal $\mathcal{M}_2^k$ is generated by the monomials $x^i y^j$ with $i + j = k$, so the submodule $\mathcal{M}_2^k \cdot \mathcal{E}_{2,2}$ is generated by the vectors $(x^i y^j, 0)$, $(0, x^i y^j)$ with $i + j = k$. Clearly, if $i \geq 1$ and $j \geq 1$ these vectors lie in T_f, so certainly in $T_f + \mathcal{M}_2^{k+1} \cdot \mathcal{E}_{2,2}$. It remains to consider $(x^k, 0)$, $(y^k, 0)$, $(0, x^k)$, $(0, y^k)$. We claim that the first two likewise lie in $T_f + \mathcal{M}_2^{k+1} \cdot \mathcal{E}_{2,2}$: when $k = 1$ this is because $(x, 0) = \frac{\partial f}{\partial y} - (0, by^{b-1})$, $(y, 0) = \frac{\partial f}{\partial x} - (0, ax^{a-1})$, and when $k \geq 2$ it is because $(x^k, 0)$, $(y^k, 0)$ are multiples of $(x^2, 0)$, $(y^2, 0)$ which appeared in our list of generators for T_f. Finally, the vector $(0, x^k)$ lies in $T_f + \mathcal{M}_2^{k+1} \cdot \mathcal{E}_{2,2}$ only when $k \geq a$, whilst the vector $(0, y^k)$ does so only when $k \geq b$. The vectors which span the required supplements are therefore the $(0, x^k)$ with $1 \leq k \leq a - 1$, and the $(0, y^k)$ with $1 \leq k \leq b - 1$. It follows that the $\mathcal{K}$-codimension of f will be $2 + (a - 1) + (b - 1) = a + b$.

Example 9 For our final illustration we shall compute the $\mathcal{K}$-codimension of the germ $f : (\mathbb{R}^2, 0) \to (\mathbb{R}^2, 0)$ given by $f(x, y) = (x^2 + y^2, x^a)$ where $a \geq 3$ is an integer. Here $\frac{\partial f}{\partial x} = (2x, ax^{a-1})$ and $\frac{\partial f}{\partial y} = (2y, 0)$ generate the Jacobian module J_f. The ideal I_f is generated by $x^2 + y^2$, x^a so the submodule $I_f \cdot \mathcal{E}_{2,2}$ is generated by $(x^2 + y^2, 0)$, $(x^a, 0)$, $(0, x^2 + y^2)$, $(0, x^a)$. Following through a computation similar to that in the previous example one finds that a supplement for the $\mathcal{K}$-tangent space in $\mathcal{M}_2 \cdot \mathcal{E}_{2,2}$ is provided by the $(0, x^i)$ and the $(0, x^{i-1}y)$ with $1 \leq i \leq a-1$, so the $\mathcal{K}$-codimension is $2 + (a - 1) + (a - 1) = 2a$.

§3. Deformations Under Contact Equivalence

The next step in our programme is to set up the basic ideas for a theory of deformations of germs under $\mathcal{K}$-equivalence: the relevance of this to the problem of classifying stable germs will be discussed in §4.

Let us start with an r-parameter deformation $F : (\mathbb{R}^r \times \mathbb{R}^m, 0) \to (\mathbb{R}^q, 0)$ of a germ $f : (\mathbb{R}^m, 0) \to (\mathbb{R}^q, 0)$. Pursuing the analogy with the Finite Dimensional Model of Chapter III one expects a major role to be played by "transversal" deformations. We need a formal interpretation for this intuitive idea, so we argue heuristically. Think of the deformation as a "germ" $(\mathbb{R}^r, 0) \to (\mathcal{E}_{m,q}, f)$ given by $u \to f_u$ where $f_u(x) = F(u, x)$. We wish this mapping to be "transverse" to the $\mathcal{K}$-orbit through f, i.e. we want something like

| image of the "differential" at 0 of this map | + | tangent space to $\mathcal{K}$-orbit through f | = | tangent space to $\mathcal{E}_{m,q}$ at f. |

The only quantity here for which we do not yet have a concrete interpretation is the "differential" at 0 of the mapping $u \to f_u$: this should be the linear mapping $\mathbb{R}^r \to \mathcal{E}_{m,q}$ which sends the standard basic vectors for $\mathbb{R}^r$ to $\frac{\partial F}{\partial u_1}\big|_{u=0}, \ldots, \frac{\partial F}{\partial u_r}\big|_{u=0}$, where we write $u_1, \ldots, u_r$ for the standard coordinates on $\mathbb{R}^r$. As a matter of convenience we write $\dot{F}_i = \frac{\partial F}{\partial u_i}\big|_{u=0}$. With this notation the image of our "differential" will be the real vector subspace $\mathbb{R}\{\dot{F}_1, \ldots, \dot{F}_r\}$ of $\mathcal{E}_{m,q}$. On this heuristic basis we introduce the following formal definition. F is a $\mathcal{K}$-transversal deformation of f when

$$\mathbb{R}\{\dot{F}_1, \ldots, \dot{F}_r\} + T_f = \mathcal{E}_{m,q}.$$

Notice therefore that f admits a $\mathcal{K}$-transversal deformation if and only if it has finite $\mathcal{K}$-codimension c, say. Assuming this to be the case one can construct explicit $\mathcal{K}$-transversal deformations by the same device used in the Finite Dimensional Model. One looks for a deformation F for which

$$F(u, x) = f(x) + u_1 f_1(x) + \ldots + u_c f_c(x) \qquad *$$

where the germs $f_1, \ldots, f_c$ are to be determined. The condition for this F to be $\overline{\mathcal{K}}$-transversal is that

$$\mathbb{R}\{f_1, \ldots, f_c\} + T_f = \mathcal{E}_{m,q}.$$

This yields an entirely practical procedure. Simply choose $f_1, \ldots, f_c$ to be a supplement to T_f, and then define F by $*$.

In practice one is interested solely in $\mathscr{K}$-transversal unfoldings of germs of rank 0. (Just why this is so will be made clear in the next section.) For such a germ f one can be a little more explicit about the form of a $\mathscr{K}$-transversal unfolding. Make the usual identification of $\mathscr{E}_{m,q}$ with the product $\mathscr{E}_m^q$: as a real vector space we can think of this as the direct sum of $\mathbb{R}^q$ with $\mathscr{M}_m^q$. We write $e_1, \ldots, e_q$ for the standard basis vectors in $\mathbb{R}^q$. Suppose f is a germ of rank 0; then a moment's thought will convince the reader that T_f is a vector subspace of $\mathscr{M}_m^q$. It will then be possible to find a basis $f_1, \ldots, f_r$ for a supplement of T_f in $\mathscr{M}_m^q$: a basis for a supplement of T_f in $\mathscr{E}_m^q$ is then provided by $e_1, \ldots, e_q, f_1, \ldots, f_r$, and we obtain a $\mathscr{K}$-transversal deformation

$$F(u, w, x) = -\sum_{i=1}^{q} w_i \cdot e_i + f(x) + \sum_{i=1}^{r} u_i \cdot f_i(x)$$

where we insert the minus sign for a minor geometric reason which will be mentioned in the next section. Here are some examples of these computations, parallel to the computations of $\mathscr{K}$-codimension given in §2, where all the work was done. In each example the germ f has rank 0.

Example 1   The germ $f : (\mathbb{R}^2, 0) \to (\mathbb{R}^2, 0)$ given by $f(x, y) = (x^2, y^2)$ has $\mathscr{K}$-codimension 4: indeed, we saw that a supplement for T_f in $\mathscr{E}_{2,2}$ is provided by $(1, 0)$, $(0, 1)$, $(y, 0)$, $(0, x)$. Thus a $\mathscr{K}$-transversal deformation is $F : (\mathbb{R}^4 \times \mathbb{R}^2, 0) \to (\mathbb{R}^2, 0)$ with components given by

$$F_1 = x^2 + u_1 y - w_1$$
$$F_2 = y^2 + u_2 x - w_2.$$

Example 2 The germ $f : (\mathbb{R}, 0) \to (\mathbb{R}^p, 0)$ defined by $f(x) = (0, \ldots, 0, x^{t+1})$, with $t \geq 1$, has $\mathscr{K}$-codimension $c = pt + p - 1$: indeed a supplement for T_f in $\mathcal{E}_{1,p}$ is provided by the $(0, \ldots, x^k, \ldots, 0)$ for which $0 \leq k \leq t$ with $(0, \ldots, 0, x^t)$ deleted. Thus a $\mathscr{K}$-transversal deformation is $F : (\mathbb{R}^c \times \mathbb{R}, 0) \to (\mathbb{R}^p, 0)$ with components given by

$$F_1 = -w_1 + u_{11}x + \ldots + u_{1,t-1}x^{t-1} + u_{1,t}x^t$$

$$F_2 = -w_2 + u_{21}x + \ldots + u_{2,t-1}x^{t-1} + u_{2,t}x^t$$

$$\vdots$$

$$F_p = -w_p + u_{p1}x + \ldots + u_{p,t-1}x^{t-1} + x^{t+1}.$$

Example 3 The germs $f : (\mathbb{R}^2, 0) \to (\mathbb{R}^2, 0)$ given by $f(x, y) = (xy, x^a \pm y^b)$, with a, b integers ≥ 3, have $\mathscr{K}$-codimension $a + b$: indeed a supplement for T_f in $\mathcal{E}_{2,2}$ is provided by the $(0, x^i)$ with $1 \leq i \leq a - 1$, and the $(0, y^i)$ with $1 \leq i \leq b - 1$, together with $(1, 0)$ and $(0, 1)$. Thus a $\mathscr{K}$-transversal deformation is the germ $F : (\mathbb{R}^{a+b} \times \mathbb{R}^2, 0) \to (\mathbb{R}^2, 0)$ with components

$$F_1 = xy - w_1$$

$$F_2 = x^a \pm y^b + \sum_{i=1}^{a-1} u_i x^i + \sum_{i=1}^{b-1} v_i y^i - w_2.$$

Example 4 The germ $f : (\mathbb{R}^2, 0) \to (\mathbb{R}^2, 0)$ given by $f(x, y) = (x^2 + y^2, x^a)$, where $a \geq 3$, has $\mathscr{K}$-codimension $2a$. A supplement for T_f in $\mathcal{E}_{2,2}$ is provided by the $(0, x^i)$ and $(0, x^{i-1}y)$ with $1 \leq i \leq a - 1$, together with $(1, 0)$ and $(0, 1)$. Thus a $\mathscr{K}$-transversal deformation is the germ $F : (\mathbb{R}^{2a} \times \mathbb{R}^2, 0) \to (\mathbb{R}^2, 0)$ with components

$$F_1 = x^2 + y^2 - w_1$$

$$F_2 = x^a + \sum_{i=1}^{a-1} u_i x^i + \sum_{i=1}^{a-1} v_i y^i - w_2 .$$

So much for examples. The next step is to pursue the analogy with the Finite Dimensional Model further to see if we can characterize the algebraic notion of "transversality" by a geometric notion of "versality". To this end we introduce a series of notions for deformations of a germ $f : (\mathbb{R}^m, 0) \to (\mathbb{R}^q, 0)$.

Equivalence of Deformations

Two r-parameter deformations F_1, F_2 of f are said to be $\underline{\mathscr{K}\text{-equivalent}}$ when there exists an r-parameter unfolding I_m of the germ at 0 of the identity mapping on $\mathbb{R}^m$ for which

$$I_m^*\bigl(F_1^*(\mathscr{M}_q)\bigr) = F_2^*(\mathscr{M}_q) .$$

In this situation we call I_m a $\underline{\mathscr{K}\text{-equivalence of deformations}}$. Of course by (2.2) this relation implies that F_1, F_2 are $\mathscr{K}$-equivalent as germs: however it says more in that the change of co-ordinates at the source has to respect the product structure on $\mathbb{R}^r \times \mathbb{R}^m$.

Induced Deformations

Suppose $F : (\mathbb{R}^r \times \mathbb{R}^m, 0) \to (\mathbb{R}^q, 0)$ is an r-parameter deformation of f, and that $H : (\mathbb{R}^s, 0) \to (\mathbb{R}^r, 0)$ is a germ. We obtain an s-parameter deformation $G : (\mathbb{R}^s \times \mathbb{R}^m, 0) \to (\mathbb{R}^q, 0)$ of f by putting

$$G(v, x) = F\bigl(H(v), x\bigr) .$$

One writes $G = H^*F$, and calls G the deformation induced by H: in this situation H is a change of parameter.

Morphisms of Deformations

Let F, G be r, s-parameter deformations of f. A morphism from F to G is a pair (H, I) with I a $\mathcal{K}$-equivalence of r-parameter deformations, and H a change of parameter, for which F is $\mathcal{K}$-equivalent to the induced deformation H^*G under I. When $r = s$ and H is invertible we refer to the morphism as an isomorphism.

Versal Deformations

A deformation G of f is $\mathcal{K}$-versal when for any deformation F there is a morphism from F to G. When f has finite $\mathcal{K}$-codimension c, say, a c-parameter $\mathcal{K}$-versal deformation is said to be $\mathcal{K}$-universal.

The main result about deformations under contact equivalence is the following analogue of (III.5.1) which we dub the $\mathcal{K}$-Versality Theorem.

(3.1)   Let $F : (\mathbb{R}^r \times \mathbb{R}^m, 0) \to (\mathbb{R}^q, 0)$ be an r-parameter deformation of a germ $f : (\mathbb{R}^m, 0) \to (\mathbb{R}^q, 0)$: a necessary and sufficient condition for F to be $\mathcal{K}$-versal is that it should be $\mathcal{K}$-transversal.

A word or two is in order concerning the analogy between this result and the Versality Theorem of Chapter III. In that result the key was the existence of neighbourhoods having a product structure, which were produced by the Inverse Function Theorem. But in the present situation the basic objects lie in a vector space of germs: one has no immediate analogue for the Inverse Function Theorem, and is forced to adopt a different stratagem. The proof that $\mathscr{K}$-transversal deformations are $\mathscr{K}$-versal is by no means easy, and we shall not give it; it uses an extension to the real case of a classical theorem of Weierstrass in complex function theory. A sketch of the result can be found in the paper of J. Martinet quoted in Appendix E. On the other hand, the converse result is relatively trivial.

Proof of Necessity

Suppose F is $\mathscr{K}$-versal. We have to show that F is $\mathscr{K}$-transversal, i.e. that

$$\mathbb{R}\{\dot{F}_1, \ldots, \dot{F}_r\} + T_f = \mathcal{E}_{m,q}.$$

Consider then a germ g in $\mathcal{E}_{m,q}$, and the 1-parameter deformation G of f given by

$$G(v, x) = f(x) + vg(x).$$

By hypothesis G is $\mathscr{K}$-equivalent (as a deformation) to an induced deformation $H = h^*F$ with $h : (\mathbb{R}, 0) \to (\mathbb{R}^r, 0)$ having components $h_1, \ldots, h_r$ say. Thus $H(t, x) = F(h(t), x)$ and

$$\dot{H} = \frac{\partial h_1}{\partial t}(0)\dot{F}_1 + \ldots + \frac{\partial h_r}{\partial t}(0)\dot{F}_r$$

establishing that $\dot{H}$ lies in $\mathbb{R}\{\dot{F}_1, \ldots, \dot{F}_r\}$. Starting from the fact that G, H are $\mathscr{K}$-equivalent deformations a computation shows that $\dot{G} - \dot{H}$ lies in T_f, and hence that $\dot{G}$ lies in $\mathbb{R}\{\dot{F}_1, \ldots, \dot{F}_r\} + T_f$. However $\dot{G} = g$, finishing the proof. □

Just as in the Finite Dimensional Model we are now in a position to justify the use of the prefix in the term "$\mathscr{K}$-universal".

3.2 **Let F, G be $\mathcal{K}$-universal deformations of a germ $f : (\mathbb{R}^m, 0) \to (\mathbb{R}^q, 0)$ of finite $\mathcal{K}$-codimension c: then F, G are $\mathcal{K}$-isomorphic deformations.**

Proof   As F is $\mathcal{K}$-versal there exists a germ $h : (\mathbb{R}^c, 0) \to (\mathbb{R}^c, 0)$ for which G is $\mathcal{K}$-equivalent to the induced deformation h^*F. That means there exists a c-parameter unfolding $\Phi : (\mathbb{R}^c \times \mathbb{R}^m, 0) \to (\mathbb{R}^c \times \mathbb{R}^m, 0)$ of the germ at 0 of the identity map on $\mathbb{R}^m$ for which $h^*F \circ \Phi$, G are $\mathcal{E}$-equivalent, i.e. $F \circ \Psi$, G are $\mathcal{E}$-equivalent, where $\Psi = (h \times 1) \circ \Phi$. We shall show that h is an invertible germ, which we do as follows. In view of (2.1) this means that there exists an invertible $q \times q$ matrix $A = A(u, x)$ with entries in $\mathcal{E}_{m+c}$ for which $F \circ \Psi = AG$, with the usual identifications. Differentiating this relation with respect to u_i, and setting $u = 0$, one obtains for $1 \leq i \leq c$ a relation of the form

$$\sum_{j=1}^{c} \frac{\partial h_j}{\partial u_i}(0) \dot{F}_j = A_0 \dot{G}_i + T_i \qquad *$$

where $A_0 = A_0(x) = A(0, x)$, and T_i lies in the $\mathcal{K}$-tangent space T_f. Now F, G are both $\mathcal{K}$-transversal deformations of f by (3.1) so $\dot{F}_1, \ldots, \dot{F}_c$ span a supplement for T_f, as do $\dot{G}_1, \ldots, \dot{G}_c$. In view of the proof of (2.4) the invertible matrix A_0 induces a linear automorphism of the $\mathcal{E}_m$-module $\mathcal{E}_m^q$ which leaves T_f invariant. Thus $A_0 \dot{G}_1, \ldots, A_0 \dot{F}_c$ also span a supplement for T_f. It follows from * that the matrix of coefficients $\frac{\partial h_j}{\partial u_i}(0)$, i.e. the Jacobian matrix of h, is invertible: the Inverse Function Theorem now enables us to deduce that h is invertible, as was required. It follows from the definition that F, G are $\overline{\mathcal{K}}$-isomorphic deformations. □

In fact we can squeeze a little more information out of the proof than is actually stated in (3.2).

(3.3) <u>Let $f : (\mathbb{R}^m, 0) \to (\mathbb{R}^q, 0)$ be a germ of finite $\mathscr{K}$-codimension c, and let F be a $\mathscr{K}$-universal deformation of f. For $d \geq c$ any d-parameter $\mathscr{K}$-versal deformation F' of f is $\mathscr{K}$-isomorphic to the $(d - c)$-parameter constant deformation of F. And hence any two d-parameter $\mathscr{K}$-versal deformations of f are $\mathscr{K}$-isomorphic.</u>

<u>Proof</u> Proceeding exactly as in (3.2) one comes to the conclusion that F' is $\mathscr{K}$-equivalent to an induced deformation h^*F, with $h : (\mathbb{R}^d, 0) \to (\mathbb{R}^c, 0)$ a submersive germ. However by (I.1.3) there exists an invertible germ $\phi : (\mathbb{R}^d, 0) \to (\mathbb{R}^d, 0)$ for which $h \circ \phi = \pi$ with $\pi : (\mathbb{R}^d, 0) \to (\mathbb{R}^c, 0)$ the projection given by $\pi(u_1, \ldots, u_d) = (u_1, \ldots, u_c)$. Thus F' is $\mathscr{K}$-isomorphic to π^*F, i.e. to the $(d - c)$-parameter constant deformation of F. The rest is clear. □

There is a small technicality which is worth mentioning at this point. We have phrased the definition of $\mathscr{K}$-isomorphism for deformations of a single germ f. One can of course phrase the definition for deformations of $\mathscr{K}$-equivalent germs f, f' and obtain results which correspond exactly to (3.2) and (3.3). We shall leave this matter to the reader, and proceed rather to the next step in our programme, which is to show how one can use the existence and uniqueness of $\mathscr{K}$-versal deformations to reduce the problem of classifying stable germs under the relation of A-equivalence to that of classifying germs of finite $\mathscr{K}$-codimension under the relation of $\mathscr{K}$-equivalence.

§4. Classification of Stable Germs

The time has come to put together the bits and gain some distance into the problem of producing explicit lists of stable germs in given dimensions. Our starting point is a very simple idea.

(4.1)   <u>Let</u> $G : (\mathbb{R}^n, 0) \to (\mathbb{R}^p, 0)$ <u>be a germ of rank</u> r: <u>then there exists an invertible germ</u> $h : (\mathbb{R}^n, 0) \to (\mathbb{R}^n, 0)$ <u>for which</u> $F = G \circ h$ <u>is an r-parameter unfolding of a germ of rank</u> 0.

<u>Proof</u>   By making linear changes of co-ordinates at source and target we can suppose that the Jacobian matrix of G, evaluated at 0, is

$$\begin{pmatrix} I_r & 0 \\ 0 & 0 \end{pmatrix}$$

where I_r is the identity $r \times r$ matrix.   Consider the germ $g : (\mathbb{R}^n, 0) \to (\mathbb{R}^r, 0)$ whose components are the first r components $G_1, \ldots, G_r$ of G. Clearly, g has rank r, so by (I.1.3) there exists an invertible germ $h : (\mathbb{R}^n, 0) \to (\mathbb{R}^n, 0)$ for which $g \circ h$ is the projection $(x_1, \ldots, x_n) \to (x_1, \ldots, x_r)$.   And then $F = G \circ h$ is the required germ. $\square$

The point of (4.1) as far as the present section is concerned is that since we are only classifying germs up to A-equivalence we can restrict our attention to r-parameter unfoldings $F : (\mathbb{R}^r \times \mathbb{R}^m, 0) \to (\mathbb{R}^r \times \mathbb{R}^q, 0)$ of rank r: for such an unfolding write $f_F : (\mathbb{R}^m, 0) \to (\mathbb{R}^q, 0)$ for the unique germ of rank 0 which F unfolds.   We wish now to study the assignation $F \to f_F$.   The first step in this direction is provided by

(4.2)   **Let $F, F' : (\mathbb{R}^r \times \mathbb{R}^m, 0) \to (\mathbb{R}^r \times \mathbb{R}^q, 0)$ be r-parameter unfoldings of germs $f_F, f_{F'}$ of rank 0: if F, F' are A-equivalent then $f_F, f_{F'}$ are $\mathcal{K}$-equivalent.**

Step 1   Since F, F' are A-equivalent there exist invertible germs H, K for which the following diagram commutes.

$$\begin{array}{ccc} (\mathbb{R}^r \times \mathbb{R}^m, 0) & \xrightarrow{F} & (\mathbb{R}^r \times \mathbb{R}^q, 0) \\ \uparrow H & & \uparrow K \\ (\mathbb{R}^r \times \mathbb{R}^m, 0) & \xrightarrow{F'} & (\mathbb{R}^r \times \mathbb{R}^q, 0) \end{array}$$

Notice first that in view of (4.1) we can suppose that $H(0, x) = (\phi(x), x)$ for some $\phi : (\mathbb{R}^m, 0) \to (\mathbb{R}^r, 0)$, and likewise that $K(0, y) = (\psi(y), y)$ for some germ $\psi : (\mathbb{R}^q, 0) \to (\mathbb{R}^r, 0)$. Write $F(u, x) = (u, f(u, x))$. It should now be clear that $f_{F'}$ is A-equivalent to the germ $g : (\mathbb{R}^m, 0) \to (\mathbb{R}^q, 0)$ given by $g(x) = f(\phi(x), x)$. And it will suffice to show that g, f_F are $\mathcal{K}$-equivalent.

Step 2 $f(u, x) - f_F(x)$ vanishes on $0 \times \mathbb{R}^m$, so each of its q components does likewise, and the Hadamard Lemma allows us to write $f(u, x) = f_F(x) + M(u, x) \cdot u$ where $M(u, x)$ is a $q \times r$ matrix whose entries are germs at 0 of functions on $\mathbb{R}^r \times \mathbb{R}^m$, and $M(0, 0) = 0$. And likewise we can write $\psi(y) = A(y) \cdot y$ where $A(y)$ is an $r \times q$ matrix whose entries are germs at 0 of functions on $\mathbb{R}^q$. It follows that we can write $\phi(x) = \psi(g(x)) = B(x) \cdot g(x)$ with B an $r \times q$ matrix. Keeping to the same matrix notation we can now write $g(x) = f_F(x) + C(x) \cdot \phi(x)$ with $C(x)$ a $q \times r$ matrix, for which $C(0) = 0$, and hence $g(x) = f_F(x) + D(x) \cdot g(x)$ with $D(x)$ a $q \times q$ matrix for which $D(0) = 0$.

This last relation we can re-write as $f_F(x) = \bigl(I - D(x)\bigr).g(x)$ where I is the identity $r \times r$ matrix. Clearly, $I - D(x)$ is an invertible matrix, and it follows from (2.1) that g, f_F are $\mathcal{C}$-equivalent, hence $\mathcal{K}$-equivalent. $\square$

It follows that the assignment $F \to f_F$ induces a mapping from A-equivalence classes of germs to $\mathcal{K}$-equivalence classes. We wish to study this mapping in detail, and to do this we need a new idea. Suppose we have an s-parameter deformation $f : (\mathbb{R}^s \times \mathbb{R}^m, 0) \to (\mathbb{R}^q, 0)$ which is submersive. A sufficiently small representative of f will be a submersion, so the zero set $V_f = f^{-1}(0)$ of f will be a smooth submanifold of $\mathbb{R}^s \times \mathbb{R}^m$ of codimension q. We take $\pi_f : (V_f, 0) \to (\mathbb{R}^s, 0)$ to be the germ at 0 of the restriction to V_f of the projection $\pi : \mathbb{R}^s \times \mathbb{R}^m \to \mathbb{R}^q$. We need the following fact.

(4.3) <u>Let</u> $f, g : (\mathbb{R}^s \times \mathbb{R}^m, 0) \to (\mathbb{R}^q, 0)$ <u>be $\mathcal{K}$-versal s-parameter deformations of germs</u> $f_0, g_0 : (\mathbb{R}^m, 0) \to (\mathbb{R}^q, 0)$ <u>of rank</u> 0: <u>if</u> f_0, g_0 <u>are $\mathcal{K}$-equivalent then</u> π_f, π_g <u>are A-equivalent</u>.

<u>Proof</u>   It is convenient to make the preliminary observation that a $\mathcal{K}$-transversal deformation (and hence any $\mathcal{K}$-versal deformation) of a germ of rank 0 is automatically submersive, so that the notation introduced above makes sense in the present context. Observe also that f, g must be $\mathcal{K}$-isomorphic deformations, in view of (3.3).

<u>Step 1</u>   Now consider first the case when $f_0 = g_0$. The fact that f, g are $\mathcal{K}$-isomorphic deformations is expressed by the existence of a commuting diagram of germs

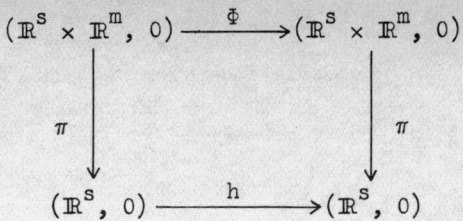

with Φ, h invertible for which $(f \circ \Phi)^*(\mathcal{M}_q) = g^*(\mathcal{M}_q)$. This last relation ensures that Φ induces a mapping from V_g onto V_f, yielding a commuting diagram of germs

$$\begin{array}{ccc} (V_g, 0) & \xrightarrow{\Phi} & (V_g, 0) \\ {\scriptstyle \pi_g} \downarrow & & \downarrow {\scriptstyle \pi_f} \\ (\mathbb{R}^s, 0) & \xrightarrow{h} & (\mathbb{R}^s, 0) \end{array}$$

expressing the fact that π_f, π_g are A-equivalent.

<u>Step 2</u>   Consider next the general case when f_0, g_0 are just $\mathcal{K}$-equivalent. By (2.1) that means that there exists an invertible germ $h : (\mathbb{R}^m, 0) \to (\mathbb{R}^m, 0)$, and an invertible $q \times q$ matrix $M(x)$ with entries in $\mathcal{E}_m$, for which $g_0(x) = M(x) \cdot f_0(h(x))$. Evidently the s-parameter deformation $g'(u, x) = M(x) \cdot f(u, h(x))$ of g_0 is $\mathcal{K}$-versal as well. It follows from Step 1 that π_g, $\pi_{g'}$ are A-equivalent. And it suffices therefore to show that π_f, $\pi_{g'}$ are A-equivalent: that however is clear as $1 \times k$ will map $V_{g'}$ onto V_f.   □

The relevance of these ideas to the material of the present section is as follows. Let $F : (\mathbb{R}^r \times \mathbb{R}^m, 0) \to (\mathbb{R}^r \times \mathbb{R}^q, 0)$ be an r-parameter unfolding of a germ f_F of rank 0, given by a formula $F(u, x) = (u, f(u, x))$. To F we associate the germ $D_F : (\mathbb{R}^r \times \mathbb{R}^q \times \mathbb{R}^m, 0) \to (\mathbb{R}^q, 0)$ given by

$(u, w, x) \to -w + f(u, x)$: thus D_F is an $(r + q)$-parameter submersive deformation of f_F. The geometric connexion between F, D_F is that V_{D_F} is just graph f, and π_{D_F} can be identified with f. The basic theorem connecting F, D_F is

(4.4) F <u>is an A-stable germ if and only if</u> D_F <u>is a $\mathcal{K}$-versal deformation of</u> f_F.

The proof of (4.4) does not use techniques lying outside the scope of this volume. However a careful version would occupy more space than is available. We shall therefore content ourselves with the statement of the result, and concentrate rather on showing the reader how one uses it to obtain explicit lists of stable germs in given dimensions. A sketch of the proof, sufficient for the competent reader, can be found in the paper of Jean Martinet mentioned in the Introduction to this book. It is perhaps worthwhile spelling out the fact that (4.4) together with the $\mathcal{K}$-Versality Theorem allows one to produce explicit examples of stable germs.

<u>Example 1</u>   We saw in Example 1 of §3 that a $\mathcal{K}$-transversal deformation of the germ (x^2, y^2) is the germ given by $(x^2 + u_1 y - w_1, y^2 + u_2 x - w_2)$. By the $\mathcal{K}$-Versality Theorem this deformation is $\mathcal{K}$-versal. It is however precisely the deformation associated to the unfolding $(u_1, u_2, x^2 + u_1 y, y^2 + u_2 x)$ so this germ must be stable.

We can now return to the main theme of this section by stating

(4.5)   <u>Let</u> $F, G : (\mathbb{R}^r \times \mathbb{R}^m, 0) \to (\mathbb{R}^r \times \mathbb{R}^q, 0)$ <u>be stable r-parameter unfoldings of germs</u> f_F, f_G <u>of rank</u> 0: <u>if</u> f_F, f_G <u>are $\mathcal{K}$-equivalent then</u> F, G <u>are A-equivalent</u>.

(4.5) is an immediate consequence of (4.3) and (4.4). In more homely language the burden of this result is that if we restrict our attention to stable germs F then the assignment $F \to f_F$ actually induces an injective mapping from A-equivalence classes f germs to $\mathscr{K}$-equivalence classes of germs. Our objective now is to determine the image of this mapping. To this end we make the following observation. If
$F : (\mathbb{R}^r \times \mathbb{R}^m, 0) \to (\mathbb{R}^r \times \mathbb{R}^q, 0)$ is a stable r-parameter unfolding of a germ $f_F : (\mathbb{R}^m, 0) \to (\mathbb{R}^q, 0)$ of rank 0 then f_F has $\mathscr{K}$-codimension $\leq r + q$: indeed (4.4) tells us that the $(r + q)$-parameter deformation D_F of f_F is $\mathscr{K}$-versal, hence $\mathscr{K}$-transversal, so the $\mathscr{K}$-codimension of f_F must be $\leq$ the number of parameters, i.e. $\leq r + q$. With this in mind write

$$S = S(r, m, q) = \text{set of } A\text{-equivalence classes of stable germs}$$
$$(\mathbb{R}^r \times \mathbb{R}^m, 0) \to (\mathbb{R}^r \times \mathbb{R}^q, 0) \text{ of rank } r.$$

$$K = K(r, m, q) = \text{set of } \mathscr{K}\text{-equivalence classes of germs}$$
$$(\mathbb{R}^m, 0) \to (\mathbb{R}^q, 0) \text{ of rank 0 and } \mathscr{K}\text{-codimension} \leq r + q.$$

We can now state the main result.

(4.6) **The mapping $S \to K$ induced by the assignment $F \to f_F$ is a bijection.**

Proof   Only surjectivity remains to be established. We consider therefore a germ $f_0 : (\mathbb{R}^m, 0) \to (\mathbb{R}^q, 0)$ of rank 0 and $\mathscr{K}$-codimension $\leq r + q$. Certainly then, as in §3, we can construct an $(r + q)$-parameter $\mathscr{K}$-transversal deformation of the form $-w + f(u, x)$ with f an r-parameter deformation of f_0. This is precisely the deformation D_F associated to the r-parameter unfolding $F : (\mathbb{R}^r \times \mathbb{R}^m, 0) \to (\mathbb{R}^r \times \mathbb{R}^q, 0)$ given by

$F(u, x) = \bigl(u, f(u, x)\bigr)$. The $\mathcal{K}$-Versality Theorem tells us that D_F is $\mathcal{K}$-versal, and then (4.4) tells us that F is stable. The trivial observation that F has rank r concludes the proof. □

Thus the problem of classifying stable germs under the relation of A-equivalence reduces to the problem of classifying germs under the relation of $\mathcal{K}$-equivalence, up to a certain $\mathcal{K}$-codimension. This latter problem we know, in principle, how to approach: it is just the analogue for map-germs of the problem for function germs discussed at length in Chapter IV. Indeed the main function of Chapter IV is to provide the reader with a model on which to base his ideas for the problem now facing us. However the mechanics of the matter are decidedly more complicated. At root we are trying to list the germs $(\mathbb{R}^m, 0) \to (\mathbb{R}^q, 0)$ of type Σ^m under the relation of $\mathcal{K}$-equivalence, at least up to a certain $\mathcal{K}$-codimension. But it may well happen that it is simply too complicated to list all the possibilities which can occur, and in order to increase one's chances of obtaining complete lists it is necessary to restrict oneself to germs defined by finer invariants than the first order symbol Σ^i. Thus the next aspect of the theory to which we address our attention is the construction of such invariants.

§5. Higher Order Singularity Sets

We have already seen that the symbol Σ^i is only a very crude invariant of a singular point. (For instance <u>any</u> singular point of a smooth function $\mathbb{R}^n \to \mathbb{R}$ of codimension ≥ 1 is of type Σ^n.) In this section we shall show how to extend the Σ^i symbolism in a natural way to obtain finer invariants. The underlying idea is maybe best understood by a detailed study of an example.

Consider the smooth mapping $\mathbb{R}^2 \to \mathbb{R}^2$ given (in complex numbers) by $z \to z^2$. Given a (small) $\epsilon > 0$ we consider the deformed map $f : \mathbb{R}^2 \to \mathbb{R}^2$ given by $z \mapsto z^2 + 2\epsilon \bar{z}$, where the bar denotes complex conjugation. In real numbers f is defined by $(x, y) \mapsto (u, v)$ where

$$u = x^2 - y^2 + 2\epsilon x \qquad v = 2xy - 2\epsilon y$$

and has Jacobian matrix

$$\begin{pmatrix} 2x + 2\epsilon & -2y \\ 2y & 2x - 2\epsilon \end{pmatrix}$$

which has rank < 2 when its determinant vanishes, i.e. on the circle $x^2 + y^2 = \epsilon^2$: this then is its singular set. The bifurcation set is soon found as well. If we parametrize the singular set by putting

$$x = \epsilon \cos \theta \qquad y = \epsilon \sin \theta$$

then we obtain a parametrization of the bifurcation set in the form

$$u = \epsilon^2 (\cos 2\theta + 2 \cos \theta) \qquad v = \epsilon^2 (\sin 2\theta - 2 \sin \theta)$$

which is a standard representation of a tricuspidal hypocycloid, the curve traced by a fixed point on a circle rolling inside another circle of three times its radius.

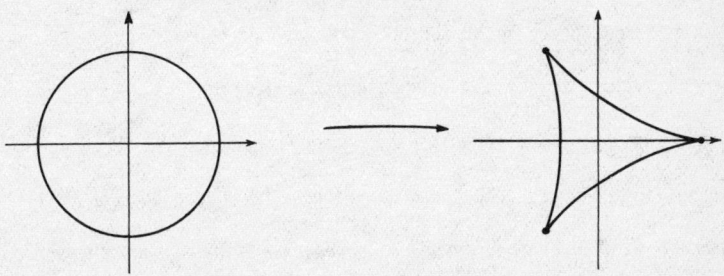

175

In fact our circle $x^2 + y^2 = \epsilon^2$ is precisely the first-order singularity set $\Sigma^1 f$, since clearly the Jacobian matrix cannot have rank 0. We are therefore unable to distinguish one point on the circle from another by just looking at the symbol Σ^i. On the other hand there are three points on the circle (the complex cube roots of ϵ^3 in fact) which very clearly need to be distinguished from the others in that they map to the cusps on the hypocycloid. A clue as to how we should distinguish these three points is obtained by further analysis.

Let us concentrate our attention on the way in which f maps the circle onto the hypocycloid. We look therefore at the restriction $f|\Sigma^1 f$. Let us compute the rank of the restriction at a point (x, y) on the circle. Recall that the differential of the restriction is the restriction of the differential of f to the tangent line to the circle. Now the tangent line to the circle at the point (x, y) is the line through the origin perpendicular to this vector. And a unit tangent vector will be $(-y/\epsilon, x/\epsilon)$. The image of this under the differential of f at (x, y) will be obtained by applying the Jacobian matrix to it, yielding the vector

$$\begin{pmatrix} 2x + 2\epsilon & -2y \\ 2y & 2x - 2\epsilon \end{pmatrix} \begin{pmatrix} -y/\epsilon \\ x/\epsilon \end{pmatrix} = \frac{2}{\epsilon} \begin{pmatrix} -2xy - \epsilon y \\ -y^2 + x^2 - \epsilon x \end{pmatrix}.$$

The differential of the restriction certainly has rank ≤ 1; and it has rank zero only when this last vector vanishes, which happens precisely at the cube roots of ϵ^3. In other words our three points are distinguished precisely by the fact that they are Σ^1 points for the restriction $f|\Sigma^1 f$, whilst all other points on the circle are Σ^0 points for the restriction.

The next step in the theory becomes clear. Given a smooth mapping $f : \mathbb{R}^n \to \mathbb{R}^p$ we have the first-order singularity sets $\Sigma^i f$. If these are submanifolds we can introduce second-order singularity sets $\Sigma^{i,j}f = \Sigma^{j}(f|\Sigma^{i}f)$. And this process can be continued. If these sets are submanifolds we can introduce third-order singularity sets $\Sigma^{i,j,k}f = \Sigma^{k}(f|\Sigma^{i,j}f)$. And so on. The sets obtained in this way are the <u>higher order Thom singularity sets</u> of f.

As Thom observed when he introduced these sets there is an unsatisfactory element here in that the definitions only make sense as long as we continue to obtain submanifolds. However, as we saw in Chapter II, there is a way out of this difficulty - at least for the first-order singularity sets. One defines submanifolds Σ^{i} of the jet-space $J^{1}(n, p)$ for which the inverse images under $j^{1}f$ are precisely the required sets $\Sigma^{i}f$: and then for a generic mapping these sets will be smooth manifolds. Thom proposed the problem of imitating this procedure for the kth order singularity sets, i.e. of defining sub-manifolds $\Sigma^{i_1,\ldots,i_k}$ in the jet-space $J^{k}(n, p)$ such that for a generic $f : \mathbb{R}^n \to \mathbb{R}^p$ the pull-back under $j^{k}f$ are smooth manifolds, and precisely the kth order singularity sets $\Sigma^{i_1,\ldots,i_k}f$. The case $k = 2$ was solved by H. Levine, but the general case waited till 1967 when it was solved by Boardman. We cannot hope to give a full account of Boardman's solution in a book of this nature: what we can do however is to describe the construction whereby one decides which symbol $\Sigma^{i_1,\ldots,i_k}$ is to be attached to the k-jet of a given map-germ, since it is an entirely practical one.

We start with an algebraic idea. Let I be a finitely generated ideal in the algebra $\mathcal{E}_n$, let $f_1, \ldots, f_p$ be generators for I, and let $y_1, \ldots, y_n$ be a system of co-ordinates in $\mathcal{E}_n$. Suppose one is given an integer $s \geq 1$. We define $\Delta_s I$ to be the ideal $I + I'$ where I' is the ideal generated by all $s \times s$ minors of the Jacobian matrix

$$\begin{pmatrix} \dfrac{\partial f_1}{\partial y_1} & \cdots & \dfrac{\partial f_1}{\partial y_n} \\ \vdots & & \vdots \\ \dfrac{\partial f_p}{\partial y_1} & \cdots & \dfrac{\partial f_p}{\partial y_n} \end{pmatrix} \qquad *$$

(5.1) <u>The ideal $\Delta_s I$ so obtained depends neither on the choice of generators, nor the choice of co-ordinates, i.e. if $g_1, \ldots, g_q$ are generators for I, and $z_1, \ldots, z_n$ is a system of co-ordinates, then $\Delta_s I$ coincides with the ideal generated by I and the $s \times s$ minors of the Jacobian matrix</u>

$$\begin{pmatrix} \dfrac{\partial g_1}{\partial z_1} & \cdots & \dfrac{\partial g_1}{\partial z_n} \\ \vdots & & \vdots \\ \dfrac{\partial g_q}{\partial z_1} & \cdots & \dfrac{\partial g_q}{\partial z_n} \end{pmatrix} \qquad **$$

<u>Proof</u>   Clearly, it suffices to show that any $s \times s$ minor of ** lies in $\Delta_s I$. Each g_i can be written as a linear combination of the f_k, with coefficients in $\mathcal{E}_n$. Thus each $\dfrac{\partial g_i}{\partial z_j}$ can be written as the same linear combination of the $\dfrac{\partial f_k}{\partial z_j}$ plus an element of I. Using the multilinearity of the determinant we see that any $s \times s$ minor of ** lies in the ideal generated by I and the $s \times s$ minors of the Jacobian matrix

$$\begin{pmatrix} \frac{\partial f_1}{\partial z_1} & \cdots & \frac{\partial f_1}{\partial z_n} \\ \vdots & & \vdots \\ \frac{\partial f_p}{\partial z_1} & \cdots & \frac{\partial f_p}{\partial z_n} \end{pmatrix} \qquad ***$$

Thus it will be enough to show that any $s \times s$ minor of *** lies in $\Delta_s I$. For this observe that the Chain Rule allows us to write each $\frac{\partial}{\partial z_j}$ as a linear combination of the $\frac{\partial}{\partial y_k}$, with coefficients in $\mathcal{E}_n$. The result now follows on appealing again to the multilinearity of the determinant. $\qquad \square$

In practice we shall work with the standard system of co-ordinates $x_1, \ldots, x_n$ but it will be important for us to know that any system will do. Note that $\Delta_s I = I$ when $s > n$, also that one has the inclusions of ideals

$$I \subseteq \Delta_n I \subseteq \Delta_{n-1} I \subseteq \cdots \subseteq \Delta_1 I.$$

<u>Example 1</u>　　Let $I = \langle x^k \rangle$ in $\mathcal{E}_1$ with $k \geq 1$. $\Delta_1 I$ is the ideal generated by I and the 1×1 minors of the Jacobian matrix (kx^{k-1}), so $\Delta_1 I = \langle x^{k-1} \rangle$. And $\Delta_s I = \langle x^k \rangle$ for $s \geq 2$.

<u>Example 2</u>　　Let $I = \langle xy, x^2 + y^2 \rangle$ in $\mathcal{E}_2$. $\Delta_1 I$ is the ideal generated by I, and the 1×1 minors of the Jacobian matrix

$$\begin{pmatrix} y & x \\ 2x & 2y \end{pmatrix}$$

so $\Delta_1 I = \langle x, y \rangle$. $\Delta_2 I$ is the ideal generated by I and the 2×2 minors; clearly $\Delta_2 I = \langle x^2, xy, y^2 \rangle$. And $\Delta_s I = \langle xy, x^2 + y^2 \rangle$ for $s \geq 3$.

Given an ideal $I \subseteq \mathcal{E}_n$ we shall adopt the notation

$$\Delta^s I = \Delta_{n-s+1} I$$

and refer to $\Delta^1 I, \Delta^2 I, \ldots, \Delta^n I$ as the successive <u>Jacobian extensions</u> of the ideal I. In view of the sequence of inclusions above we have

$$I = \Delta^0 I \subseteq \Delta^1 I \subseteq \Delta^2 I \subseteq \ldots \subseteq \Delta^n I . \qquad *$$

Let us call I <u>proper</u> when $I \neq \mathcal{E}_n$. (Note: $I = \mathcal{E}_n$ is the same thing as saying that if we take a finite set of generators for I then at least one generator has constant term $\neq 0$.) Suppose I is proper. The <u>critical Jacobian extension</u> of I is the last ideal $\Delta^{i_1} I$ in the sequence * which is proper. It has in turn a critical Jacobian extension $\Delta^{i_2} \Delta^{i_1} I$. And so on. In this way we obtain an ascending sequence $\Delta^{i_1} I, \Delta^{i_2} \Delta^{i_1} I, \ldots$ of successive critical Jacobian extension of I, and we say that I has <u>Boardman symbol</u> $(i_1, i_2, \ldots)$.

Example 3   One checks easily that the Boardman symbol of the ideal $I = \langle x^k \rangle$ in $\mathcal{E}_1$ mentioned in Example 1 is $(1, 1, \ldots, 1, 0, \ldots)$ with $(k-1)$ repeated 1's. And the Boardman symbol of the ideal $I = \langle xy, x^2 + y^2 \rangle$ in $\mathcal{E}_2$ mentioned in Example 2 is $(2, 0, 0, \ldots)$.

The <u>Boardman symbol</u> of a germ $f : (\mathbb{R}^n, 0) \to (\mathbb{R}^p, 0)$ is defined to be that of the ideal I_f generated by the components $f_1, \ldots, f_p$.

(5.2)   <u>The Boardman symbol of a germ $(\mathbb{R}^n, 0) \to (\mathbb{R}^p, 0)$ is a contact invariant, i.e. if two such germs are $\mathcal{K}$-equivalent they have the same Boardman symbol</u>.

Proof   Suppose the germs are f with components $f_1, \ldots, f_p$ and f' with components $f'_1, \ldots, f'_p$.

Step 1   Suppose f, f' are $\mathscr{C}$-equivalent. By (2.1) the ideals I_f, $I_{f'}$ coincide, so the Boardman symbols of f, f' coincide, by (5.1).

Step 2   Suppose f, f' are right-equivalent, i.e. there exists an invertible germ $h : (\mathbb{R}^n, 0) \to (\mathbb{R}^n, 0)$ for which $f \circ h, f'$ coincide. The components $h_1, \ldots, h_n$ of h yield a system of co-ordinates. If we compute the Jacobian matrix of $f'_1, \ldots, f'_p$ relative to the standard system of co-ordinates $x_1, \ldots, x_n$ we get the same ideal as if we compute the Jacobian matrix of $f_1, \ldots, f_p$ relative to the system of co-ordinates $h_1, \ldots, h_n$. Again it follows from (5.1) that f, f' have the same Boardman symbol.

Step 3   The required result is immediate from the two preceding steps. $\square$

Now we can extend our definition. Certainly any germ $f : (\mathbb{R}^n, x) \to (\mathbb{R}^p, y)$ is $\mathscr{K}$-equivalent to a germ $f_0 : (\mathbb{R}^n, 0) \to (\mathbb{R}^p, 0)$. We define the <u>Boardman symbol</u> of f to be that of f_0. In view of (5.2) this definition is unambiguous.

(5.3)   <u>The first k integers in the Boardman symbol of a germ $f : (\mathbb{R}^n, x) \to (\mathbb{R}^p, y)$ depend only on the k-jet of f.</u>

Proof   Clearly, we can suppose $x = 0$, $y = 0$. Let I be the ideal generated by the components $f_1, \ldots, f_p$ of f, and let $(i_1, i_2, \ldots)$ be the Boardman symbol of I. It is evident, by induction on k, that an ideal $\Delta^{i_s} \Delta^{i_{k-1}} \ldots \Delta^{i_1} I$ is generated by partial derivatives of order $\leq k$ of $f_1, \ldots, f_p$. And whether this ideal is proper or not depends only on the values of all these derivatives at 0: thus i_k depends only on the k-jet of f. □

We return now to our objective of partitioning the jet-space $J^k(n, p)$ into k^{th} order singularity sets. Given k integers $i_1, \ldots, i_k$ we say that a germ $f : (\mathbb{R}^n, x) \to (\mathbb{R}^p, y)$ is of <u>type</u> $\Sigma^{i_k,\ldots,i_k}$ when its Boardman symbol has the form $(i_1, \ldots, i_k; \ldots)$. We define $\Sigma^{i_1,\ldots,i_k}$ to be the subset of the jet-space comprising those jets having a representative germ of type $\Sigma^{i_1,\ldots,i_k}$: in view of (5.3) this definition is unambiguous.

<u>Example 4</u> In the case k = 1 the above definition recovers the first-order singularity sets studied in Chapter II. One sees this as follows. Consider a jet in $J^1(n, p)$ having a representative germ with zero source and target, and let I be the ideal generated by the components of the germ. Now $\Delta^s I$ is generated by I and the minors of order (n - s + 1) of the Jacobian matrix, and will be proper if and only if all the minors of order (n - s + 1) are zero, i.e. if and only if the Jacobian has kernel rank ⩾ s: it follows that $\Delta^s I$ will be critical if and only if the Jacobian has kernel rank precisely s - which is the same thing as saying that the jet lies in the first-order singularity set Σ^s, as defined in Chapter II.

Before we turn to further examples we shall determine just when the kth order singularity set $\Sigma^{i_1,\ldots,i_k}$ in $J^k(n, p)$ is non-empty. The answer is provided by

(5.4)   <u>A necessary and sufficient condition for the set</u> $\Sigma^{i_1,\ldots,i_k} \subseteq J^k(n, p)$ <u>to be non-empty is that the following conditions should be satisfied:</u>

(i) $n \geq i_1 \geq i_2 \geq \ldots \geq i_k \geq 0$

(ii) $i_1 \geq n - p$

(iii) if $i_1 = n - p$ then $i_1 = i_2 = \ldots = i_k$.

<u>Proof</u>   Note first that we need only concern ourselves with jets having zero source and target.

<u>Necessity</u>   (i) It is an immediate consequence of the definitions that $n \geq i_1$, and that $i_1, i_2, \ldots, i_k$ are all ≥ 0. To see that $i_j \geq i_{j+1}$ we proceed as follows. Take I to be the ideal in $\mathcal{E}_n$ generated by the components of a representative of some jet in $\Sigma^{i_1, \ldots, i_k}$. We can suppose that $\Delta^{i_t} \ldots \Delta^{i_1} I$ is generated by $g_1, \ldots, g_{\sigma_t}$ say, with $\sigma_1 \leq \sigma_2 \leq \ldots$. Suppose $\Delta^s \Delta^{i_j} \ldots \Delta^{i_1} I$ is proper. This implies that all the minors of order $(n - s + 1)$ of the Jacobian matrix of $g_1, \ldots, g_{\sigma_j}$ vanish, and therefore in particular all the minors of order $(n - s + 1)$ of the Jacobian matrix of $g_1, \ldots, g_{\sigma_{j-1}}$ vanish, so that the ideal $\Delta^s \Delta^{i_{j-1}} \ldots \Delta^{i_1} I$ is proper, and hence $s \leq i_j$. Taking $s = i_{j+1}$ we obtain $i_{j+1} \leq i_j$.

(ii) As we observed above in Example 4 the first index i_1 is the kernel rank of the jet, so certainly $\geq n - p$.

(iii) If $i_1 = n - p$ then $\Delta^{i_1} I = \Delta_{p+1} I = I$, and it follows immediately that $i_1 = i_2 = \ldots = i_k$.

<u>Sufficiency</u>   Suppose conditions (i), (ii) and (iii) are satisfied. We have to produce a germ $f : (\mathbb{R}^n, 0) \to (\mathbb{R}^p, 0)$ with components $f_1, \ldots, f_p$ say, which is of type $\Sigma^{i_1, \ldots, i_k}$. We consider cases, leaving the computations as good exercises for the reader.

<u>Case when</u> $i_1 = n - p$ In this case we choose $f_1 = x_1, \ldots, f_p = x_p$.

<u>Case when</u> $i_1 > n - p$ In this case the following choice will do.

$$\begin{cases} f_i = x_i & (1 \leq i \leq n - i_1) \\ f_{n-i_1+1} = \sum_{n-i_1+1}^{n-i_1} x_j^2 + \sum_{n-i_2+1}^{n-i_3} x_j^3 + \cdots \\ f_i = 0 & (n - i_1 + 2 \leq i \leq p) \end{cases}$$

$\square$

One immediate consequence of this result is that the partition of $J^k(n, p)$ by the non-empty $\Sigma^{i_1, \ldots, i_k}$ is actually finite.

<u>Example 5</u>   The only non-empty singularity sets in the jet-space $J^2(2, 2)$ are $\Sigma^{2,2}$, $\Sigma^{2,1}$, $\Sigma^{2,0}$, $\Sigma^{1,1}$, $\Sigma^{1,0}$ and $\Sigma^{0,0}$.

<u>Example 6</u>   Let us return to the example we studied in detail at the beginning of this section, i.e. the mapping $f : \mathbb{R}^2 \to \mathbb{R}^2$ defined by $(x, y) \to (u, v)$ with $u = x^2 - y^2 + 2\epsilon x$, $v = 2xy - 2\epsilon y$ and $\epsilon > 0$. We shall compute the second-order Boardman symbol $\Sigma^{i,j}$ for the germ of f at any point (x_0, y_0) in the plane. By definition, we need to compute the Boardman symbol of any germ $f_0 : (\mathbb{R}^2, 0) \to (\mathbb{R}^2, 0)$ which is $\mathscr{K}$-equivalent to the germ of f at (x_0, y_0). An obvious choice for f_0 is given by $(x, y) \mapsto (u_0, v_0)$ where

$$\begin{cases} u_0(x, y) = u(x + x_0, y + y_0) - u(x_0, y_0) \\ v_0(x, y) = v(x + x_0, y + y_0) - v(x_0, y_0) \end{cases}.$$

Here, I is the ideal generated by u_0, v_0 and $\Delta^i I$ is the ideal generated by u_0, v_0 and the minors of order $(3 - i)$ of their Jacobian matrix

$$\begin{pmatrix} \dfrac{\partial u_0}{\partial x} & \dfrac{\partial u_0}{\partial y} \\ \\ \dfrac{\partial v_0}{\partial x} & \dfrac{\partial v_0}{\partial y} \end{pmatrix}.$$

The ideal $\Delta^2 I$ is generated by u_0, v_0 and the entries in the Jacobian, and cannot be proper as two of its generators $\dfrac{\partial u_0}{\partial x}$, $\dfrac{\partial v_0}{\partial y}$ have constant term $\neq 0$. The ideal $\Delta^1 I$ is generated by u_0, v_0 and the determinant D of the above Jacobian, and will be proper (hence critical) when the constant term $x_0^2 + y_0^2 - \epsilon^2$ in D vanishes: this then, as we saw before, is the set where the germ of f has type Σ^1. The ideal $\Delta^1 \Delta^1 I$ is generated by u_0, v_0, D and the 2×2 minors of their Jacobian: a line or two of computation will verify that the constant terms in these generators are

$$x_0^2 + y_0^2 - \epsilon^2; \quad y_0(2x_0 + \epsilon); \quad y_0^2 - x_0^2 + x_0 \epsilon.$$

And the ideal $\Delta^1 \Delta^1 I$ will be critical exactly when the last three expressions vanish simultaneously, i.e. exactly at the three complex cube roots of ϵ^3. Thus the three exceptional points on the circle $\Sigma^1 f$ are distinguished precisely by the fact that the germ of f at these points has type $\Sigma^{1,1}$, whereas at all other points on the circle it has type $\Sigma^{1,0}$.

In Chapter II we proved that the first-order singularity sets $\Sigma^{i_1} \subseteq J^1(n, p)$ were smooth manifolds, and computed their codimensions. We shall only state the much harder result of Boardman.

(5.5) <u>If the kth order singularity set $\Sigma^{i_1,\ldots,i_k}$ in $J^k(n, p)$ is non-empty then it is a smooth submanifold of codimension</u>

$$(p - n + i_1)\mu(i_1, \ldots, i_k) - (i_1 - i_2)\mu(i_2, \ldots, i_k) - \ldots - (i_{k-1} - i_k)\mu(i_k)$$

where $\mu(i_1, \ldots, i_k)$ <u>denotes the number of sequences</u> $(j_1, \ldots, j_k)$ <u>of</u> <u>integers which satisfy the following conditions</u>

(i) $j_1 \geq j_2 \geq \ldots \geq j_k \geq 0$

(ii) $i_s \geq j_s$ for all $1 \leq s \leq k$ and $j_1 > 0$.

<u>Example 7</u>   In the case $k = 1$ we have $\mu(i) = i$ and hence the codimension of Σ^i in $J^1(\mathbb{R}^n, \mathbb{R}^p)$ is $(p - n + i)i$, which agrees with the formula we obtained in Chapter II.

<u>Example 8</u>   Suppose $i_1 = i_2 = \ldots = i_k = 1$. Clearly we have $\mu(1, \ldots, 1) = k$, so the codimension of $\Sigma^{1,\ldots,1}$ in $J^k(n, p)$ will be $(p - n + 1)k$. Note that in the equidimensional case $p = n$ the answer is just k, the number of repeated 1's.

<u>Example 9</u>   In the case $k = 2$ one has $\mu(i, j) = i(j + 1) - \frac{j(j-1)}{2}$, so the codimension of $\Sigma^{i,j}$ in $J^2(n, p)$ is given by the formula

$$(p - n + i)i + \tfrac{j}{2}[(p - n + i)(2i - j + 1) - 2i + 2j].$$

In view of (5.5) the singularity sets $\Sigma^{i_1,\ldots,i_k}$ are called the <u>Boardman</u> <u>submanifolds</u> of $J^k(n, p)$. The Thom Transversality Lemma, proved in §4 of Chapter II, yields the following.

(5.6)   <u>The set of all smooth mappings</u> $f : \mathbb{R}^n \to \mathbb{R}^p$ <u>for which</u> $j^k f$ <u>is</u> <u>transverse to all the Boardman submanifolds</u> $\Sigma^{i_1,\ldots,i_k}$ <u>is dense in</u> $C^\infty(\mathbb{R}^n, \mathbb{R}^p)$.

We shall call a smooth mapping $f : \mathbb{R}^n \to \mathbb{R}^p$ <u>generic in the sense of Boardman</u> when $j^k f$ is transverse to all the Boardman submanifolds $\Sigma^{i_1,\ldots,i_k}$, for every integer $k \geq 1$. For such a mapping the set

$$\Sigma^{i_1,\ldots,i_k} f = (j^k f)^{-1}(\Sigma^{i_1,\ldots,i_k})$$

will be a smooth submanifold of $\mathbb{R}^n$ having the same codimension as $\Sigma^{i_1,\ldots,i_k}$. Boardman showed that

(5.7)   <u>Let $f : \mathbb{R}^n \to \mathbb{R}^p$ be generic in the sense of Boardman: then</u>
$$\Sigma^{i_1,\ldots,i_k,i_{k+1}} f = \Sigma^{i_{k+1}}(f | \Sigma^{i_1,\ldots,i_k} f).$$

In other words we have the following. Any smooth mapping $f : \mathbb{R}^n \to \mathbb{R}^p$ can be forced to be generic in the sense of Boardman by an arbitrarily small perturbation: moreover, for such a mapping the sets $\Sigma^{i_1,\ldots,i_k} f$ are smooth manifolds and coincide precisely with the Thom singularity sets. Note incidentally one trivial consequence of (5.7), namely that

$$\Sigma^{i_1} f \supseteq \Sigma^{i_1,i_2} f \supseteq \Sigma^{i_1,i_2,i_3} f \supseteq \ldots$$

<u>Example 10</u>   Let $f : \mathbb{R}^3 \to \mathbb{R}^3$ be generic in the sense of Boardman. We ask which Thom singularity sets can occur. By Example 7 the codimension of Σ^i in $J^1(3, 3)$ is i^2, hence $\Sigma^i f$ has codimension i^2 in $\mathbb{R}^3$: clearly then $\Sigma^1 f$ with codimension 1 is the only first-order Thom singularity set which can occur. $\Sigma^1 f$ splits into $\Sigma^{1,0} f$ and $\Sigma^{1,1} f$ with codimensions 1, 2 respectively, and $\Sigma^{1,1} f$ splits into $\Sigma^{1,1,0} f$ and $\Sigma^{1,1,1} f$ with codimensions 2, 3 respectively; no further splitting can take place since the kth order Thom singularity set $\Sigma^{1,\ldots,1} f$ has codimension k, so will not appear for $k \geq 4$.

Example 11   A pleasant illustration is provided by the dovetail mapping $f : \mathbb{R}^3 \to \mathbb{R}^3$ given by $(x, y, z) \mapsto (u, v, w)$ where

$$u = x \quad : \quad v = y \quad : \quad w = z^4 - uz - vz^2.$$

The reader will easily check that the possible Thom singularity sets $\Sigma^{1,\ldots,1} f$ are given by the equations below.

$$\Sigma^1 f \; : \; \frac{\partial w}{\partial z} = 0$$

$$\Sigma^{1,1} f \; : \; \frac{\partial w}{\partial z} = 0 \quad \text{and} \quad \frac{\partial^2 w}{\partial z^2} = 0$$

$$\Sigma^{1,1,1} f \; : \; \frac{\partial w}{\partial z} = 0 \quad \text{and} \quad \frac{\partial^2 w}{\partial z^2} = 0 \quad \text{and} \quad \frac{\partial^3 w}{\partial z^3} = 0.$$

$\Sigma^1 f$ is the folded surface illustrated below: $\Sigma^{1,1} f$ is the fold curve, and $\Sigma^{1,1,1} f$ is the origin.

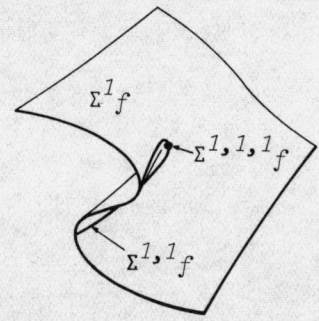

In succeeding sections it will be useful for us to know that the Boardman symbols $\Sigma^{i_1,\ldots,i_k}$ are invariant under unfolding in the following precise sense.

(5.8)   Let $F : (\mathbb{R}^r \times \mathbb{R}^n, 0) \to (\mathbb{R}^r \times \mathbb{R}^p, 0)$ be an r-parameter unfolding of the germ $f : (\mathbb{R}^n, 0) \to (\mathbb{R}^p, 0)$: then f, F have the same Boardman symbol.

Proof   Let $x_1, \ldots, x_n$ be the standard co-ordinates in $\mathbb{R}^n$, and $u_1, \ldots, u_r$ those in $\mathbb{R}^r$. We take f' to be the germ whose components are $u_1, \ldots, u_r, f_1, \ldots, f_p$ where $f_1, \ldots, f_p$ are the components of f. The proof proceeds in two steps.

Step 1   We claim first that f, f' have the same Boardman symbol. Suppose f has Boardman symbol $(i_1, i_2, \ldots)$. We shall show, by induction on k, that the ideal $\Delta^s \Delta^{i_{k-1}} \ldots \Delta^{i_0} I_{f'}$ is generated by $u_1, \ldots, u_r$ and $\Delta^s \Delta^{i_{k-1}} \ldots \Delta^{i_0} I_f$. Here we put $i_0 = 0$ for convenience, and tacitly identify $\mathcal{E}_n$ with a subset of $\mathcal{E}_{n'}$ where $n' = n + r$. It will follow immediately that $\Delta^{i_k} \ldots \Delta^{i_0} I_{f'}$ is critical, so establishing the claim. When $k = 0$ the assertion is trivial. Suppose it holds for k. Take J to be the Jacobian matrix of some fixed set of generators for $\Delta^{i_k} \ldots \Delta^{i_0} I_f$ relative to the standard co-ordinates $x_1, \ldots, x_n$, and J' to be the Jacobian matrix of the same set of generators augmented by $u_1, \ldots, u_r$ relative to the co-ordinates $x_1, \ldots, x_n, u_1, \ldots, u_r$. Consider the ideal $\Delta^s \Delta^{i_k} \ldots \Delta^{i_0} I_{f'}$ generated by $u_1, \ldots, u_r$, the generators for $\Delta^{i_k} \ldots \Delta^{i_0} I_f$, and the minors of order $(n' - s + 1)$ of J'. Now J' is the direct sum of J and the identity $r \times r$ matrix, so the ideal generated by the minors of order $(n' - s + 1)$ of J' coincides with the ideal generated by the minors of order $(n - s + 1)$ of J. Thus $\Delta^s \Delta^{i_k} \ldots \Delta^{i_0} I_{f'}$ is generated by $u_1, \ldots, u_r$, the generators for $\Delta^{i_k} \ldots \Delta^{i_1} I_f$, and the minors of order $(n - s + 1)$ of J, i.e. it is generated by $u_1, \ldots, u_r$ and $\Delta^s \Delta^{i_k} \ldots \Delta^{i_0} I_f$.

Step 2   We claim that f', F are $\mathcal{E}$-equivalent, and hence $\mathcal{K}$-equivalent. It will then follow from (5.2) that f', F have the same Boardman symbol, which fact combined with Step 1 will clinch the result. Since F

unfolds f it has components $u_1, \ldots, u_r, F_1, \ldots, F_p$ and for $1 \leq i \leq p$ we have

$$F_i(x_1, \ldots, x_n, 0, \ldots, 0) = f_i(x_1, \ldots, x_n)$$

identically in $x_1, \ldots, x_n$. It follows from the Hadamard Lemma that we can write each $F_i = f_i + \zeta_i$ where ζ_i lies in the ideal in $\mathcal{E}_n$, generated by $u_1, \ldots, u_r$. Thus $u_1, \ldots, u_r, f_1, \ldots, f_p$ and $u_1, \ldots, u_r, F_1, \ldots, F_p$ generate the same ideal in $\mathcal{E}_{n'}$, so the corresponding germs f', F are $\mathcal{E}$-equivalent by (2.1). □

Example 12   The germ $F : (\mathbb{R}^n, 0) \to (\mathbb{R}^n, 0)$ with the components $F_1, \ldots, F_n$ where

$$\begin{cases} F_i = x_i & (1 \leq i \leq n - 1) \\ F_n = x_n^{n+1} + \sum_{i=1}^{n-1} x_i x_n^i \end{cases}$$

is an unfolding of the germ $f : (\mathbb{R}, 0) \to (\mathbb{R}, 0)$ given by $f(x) = x^{n+1}$ so has type $\Sigma^{1,\ldots,1,0}$ with n repeated 1's using Example 3.

Example 13   The germ $F : (\mathbb{R}^4, 0) \to (\mathbb{R}^4, 0)$ with the components F_1, F_2, F_3, F_4 where

$$\begin{cases} F_1 = x_1 \\ F_2 = x_2 \\ F_3 = x_3 x_4 \\ F_4 = x_3^2 + x_4^2 + x_1 x_3 + x_2 x_4 \end{cases}$$

is an unfolding of the germ $f : (\mathbb{R}^2, 0) \to (\mathbb{R}^2, 0)$ given by $f(x, y) = (xy, x^2 + y^2)$ so has type $\Sigma^{2,0}$ by Example 3.

§6. Classifying Germs under $\mathcal{K}$-equivalence

In this section we shall consider the very simplest situations where it is possible to obtain explicit lists of germs under the relation of $\mathcal{K}$-equivalence. As in Chapter IV the whole thing turns on the idea of "determinacy". We call a germ $f : (\mathbb{R}^n, 0) \to (\mathbb{R}^p, 0)$ $\underline{\mathcal{K}\text{-k-determined}}$ when any germ $g : (\mathbb{R}^n, 0) \to (\mathbb{R}^p, 0)$ with $j^k f = j^k g$ is $\mathcal{K}$-equivalent to f. By analogy with (IV.3.1) one might reasonably expect the following result.

(6.1) $\underline{\text{A sufficient condition for a germ}}$ $f : (\mathbb{R}^n, 0) \to (\mathbb{R}^p, 0)$ $\underline{\text{to be}}$ $\underline{\mathcal{K}\text{-k-determined is that}}$ $\mathcal{M}_n^{k+1} \cdot \mathcal{E}_{n,p} \subseteq T_f$.

In fact the result is correct, and its proof turns out to be no more than a slightly complicated version of the proof of (IV.3.1): for that reason we shall omit the proof. By further analogy with Chapter IV we call f $\underline{\text{finitely }\mathcal{K}\text{-determined}}$ when it is $\mathcal{K}$-k-determined for some $k \geq 1$. It follows immediately from (2.3) and (6.1) that a germ of finite $\mathcal{K}$-codimension must be finitely $\mathcal{K}$-determined. In fact that statement is the only application we shall make of (6.1) in this book.

As a starting point, let us look again at the case of germs of functions, only this time under $\mathcal{K}$-equivalence rather than $\mathcal{R}$-equivalence. The first step is the Splitting Lemma.

(6.2) Let $f \in \mathscr{M}_n^2$ be a germ of corank c and finite $\mathscr{K}$-codimension: then f is $\mathscr{K}$-equivalent to a germ
$$g(x_1, \ldots, x_c) \pm x_{c+1}^2 \pm \ldots \pm x_n^2$$
with $g \in \mathscr{M}_c^3$.

The proof is exactly the same as that given in Chapter IV. The normal forms for germs of corank 0 are given by the Morse Lemma. And for germs of corank 1 one obtains almost exactly the same classification as before.

(6.3) Let $f \in \mathscr{M}_n^2$ have corank 1 and finite $\mathscr{K}$-codimension $k \geq 1$: then f is $\mathscr{K}$-equivalent to a germ of the form $x_1^{k+1} \pm x_2^2 \pm \ldots \pm x_n^2$.

Again, the proof is exactly the same as that of the corresponding result in Chapter IV. One can continue in this way, just as we did in Chapter IV, and it turns out that up to a certain point the two classifications are more or less identical, but then they begin to diverge. We shall not pursue the point further.

Let us now turn our attention to germs of smooth mappings, as opposed to germs of smooth functions. We are now in a much more complicated situation, and can only hope to obtain results in very special cases. It is in just this situation that the Boardman symbol of a germ proves to be useful, in that it allows us to distinguish special cases. The first fact one ought to be aware of is

(6.4) Let $f : (\mathbb{R}^n, 0) \to (\mathbb{R}^p, 0)$ be a germ of finite $\mathscr{K}$-codimension: its Boardman symbol must have the form $(i_1, \ldots, i_k, 0, 0, \ldots)$ for some integer $k \geq 1$.

Proof   As we have remarked already f must be finitely $\mathscr{K}$-determined so $\mathscr{K}$-equivalent to $g : (\mathbb{R}^n, 0) \to (\mathbb{R}^p, 0)$ each of whose components is given by a polynomial of degree $\le k$, for some integer $k \ge 1$. Clearly, the Boardman symbol of g has the required form $(u_1, \ldots, u_k, 0, \ldots, 0)$, and since the Boardman symbol is a $\mathscr{K}$-invariant, that of f has the same form. □

Perhaps the simplest case to study for germs $(\mathbb{R}^n, 0) \to (\mathbb{R}^p, 0)$ with $p \ge 2$ is the case when $n = 1$. The first order Boardman symbol is Σ^0 or Σ^1. In the former case the germ is non-singular and a normal form is provided by (I.1.4). Now suppose the germ has finite $\mathscr{K}$-codimension, and type Σ^1. We ask for the classification up to $\mathscr{K}$-equivalence.

(6.5)   <u>Let $f : (\mathbb{R}, 0) \to (\mathbb{R}^p, 0)$ be a germ of type Σ^1, and finite $\mathscr{K}$-codimension. Then f is necessarily of type $\Sigma^{1,\ldots,1,0}$</u> (with k <u>repetitions</u>) <u>for some integer $k \ge 1$, and in that case is $\mathscr{K}$-equivalent to</u> $(0, 0, \ldots, 0, x^{k+1})$.

Proof   That f has the type indicated is an immediate consequence of (5.4) and (6.4). Next, we claim that f has type $\Sigma^{1,\ldots,1,0}$ (with k repetitions) if and only if the following conditions are satisfied.

(i) $\dfrac{\partial^j f_1}{\partial x^j}(0) = 0, \ldots, \dfrac{\partial^j f_p}{\partial x^j}(0) = 0$ $(j \le k)$

(ii)   some $\dfrac{\partial^j f_i}{\partial x^j}(0) \ne 0$ $(j = k + 1)$

where $f_1, \ldots, f_p$ denote the components of f. Indeed by induction on k one readily checks that $\Delta^1 \ldots \Delta^1 I_f$ is generated by I_f, and the derivatives $\dfrac{\partial^j f_i}{\partial x^j}$ with $1 \le i \le p$, $1 \le j \le k$. The claim follows immediately.

In view of (IV.2.4) the conditions (i) and (ii) are equivalent to saying that $I_f = \langle x^{k+1} \rangle$, and then (2.1) tells us that this is the same as f being $\mathscr{E}$-equivalent to $(0, 0, \ldots, x^{k+1})$. The result follows. □

So far so good. Let us press on further. If we insist on both n and p being ≥ 2, then the next simplest case to study is that of germs $f : (\mathbb{R}^2, 0) \to (\mathbb{R}^2, 0)$. Here the possible first-order Boardman symbols are Σ^0, Σ^1, Σ^2. In the first case the germ is non-singular, and a normal form is provided by (I.1.3). For germs of type Σ^1 we have the following result.

(6.6)   <u>Let $f : (\mathbb{R}^2, 0) \to (\mathbb{R}^2, 0)$ be a germ of type Σ^1, and finite $\mathscr{K}$-codimension. Then f is necessarily of type $\Sigma^{1,\ldots,1,0}$ (with k repetitions) for some integer $k \geq 1$, and in that case is $\mathscr{K}$-equivalent to (x, y^{k+1}).</u>

<u>Proof</u>   As in the proof of the preceding proposition, the fact that f has the type indicated is an immediate consequence of (5.4) and (6.4). As f has rank 1 we know from (4.1) that f can be assumed to be a 1-parameter unfolding of a germ $f_0 : (\mathbb{R}, 0) \to (\mathbb{R}, 0)$ of rank 0, which by (5.8) also has type $\Sigma^{1,\ldots,1,0}$ (with k repetitions). And clearly f is then $\mathscr{K}$-equivalent to $(x, f_0(y))$. But (6.5) tells us that $f_0(y)$ is $\mathscr{K}$-equivalent to y^{k+1}, and hence f is $\mathscr{K}$-equivalent to (x, y^{k+1}). □

We have still to treat germs $(\mathbb{R}^2, 0) \to (\mathbb{R}^2, 0)$ of type Σ^2. At this point we are entering a decidedly more complex situation, and must proceed cautiously. The possible second-order Boardman symbols are $\Sigma^{2,0}$, $\Sigma^{2,1}$, $\Sigma^{2,2}$ in order of increasing degeneracy. For the first we can obtain a complete result (due to J. Mather).

(6.7) **Any germ** $f : (\mathbb{R}^2, 0) \to (\mathbb{R}^2, 0)$ **of type** $\Sigma^{2,0}$ **and finite $\mathcal{K}$-codimension is $\mathcal{K}$-equivalent to one of the germs listed below.**

$$I_{a,b} : (xy, x^a + y^b) \qquad b \geq a \geq 2$$

$$II_{a,b} : (xy, x^a - y^b) \qquad b \geq a \geq 2, \; a \text{ even}$$

$$IV_a : (x^2 + y^2, x^a) \qquad a \geq 3$$

Note   We have kept to Mather's notation $I_{a,b}$, $II_{a,b}$, IV_a for these germs. His list, which was for a more general situation, included in addition certain germs denoted $III_{a,b}$, V_a.

Proof   The first thing to note is that the components of f have no linear terms, so the 2-jet of f can be thought of as a pair of binary quadratic forms

$$(a_1 x^2 + 2b_1 xy + c_1 y^2, \; a_2 x^2 + 2b_2 xy + c_2 y^2)$$

In Chapter III we saw that by applying <u>linear</u> changes of co-ordinates at source and target we can suppose that the 2-jet has one of the normal forms written out in the table below. Beside each normal form we have written the second-order Boardman symbol. The computations are very easy, and left to the reader.

normal form for pencil	Boardman symbol
$(xy,\ x^2 \pm y^2)$	$\Sigma^{2,0}$
$(xy,\ x^2)$	$\Sigma^{2,0}$
$(xy,\ 0)$	$\Sigma^{2,0}$
$(x^2 + y^2,\ 0)$	$\Sigma^{2,0}$
$(x^2,\ 0)$	$\Sigma^{2,1}$
$(0,\ 0)$	$\Sigma^{2,2}$

Since the Boardman symbol is a $\mathscr{K}$-invariant we can discard the last two normal forms in the table. Certainly then the first component of f will have 2-jet xy or $x^2 + y^2$. We consider these cases separately.

<u>The xy case</u>   The first component of f is a smooth germ $(\mathbb{R}^2, 0) \to (\mathbb{R}, 0)$ of corank 0. It follows from the Morse Lemma that this germ is $\mathscr{R}$-equivalent to xy. Applying the same change of co-ordinates to f we see that it is $\mathscr{K}$-equivalent to a germ $\bigl(xy, \zeta(x, y)\bigr)$ where ζ has no linear terms. We have supposed f is of finite $\mathscr{K}$-codimension, so finitely $\mathscr{K}$-determined, and therefore $\zeta(x, y)$ can be supposed to be a polynomial. It follows from (2.1) that we shall not change the $\mathscr{K}$-equivalence class by subtracting from ζ a multiple of xy in $\mathcal{E}_2$: thus we can suppose our germ has the form $\bigl(xy, \alpha(x) + \beta(y)\bigr)$ with α, β polynomials.

At this point a couple of remarks are in order. Suppose $\alpha \neq 0$, and has order $a \geq 2$ (i.e. x^a is the lowest power of x which appears in α): then certainly we can produce a change of co-ordinates $x \mapsto X$ under which α becomes $\pm x^a$. And similarly, if $\beta \neq 0$ and has order $b \geq 2$ we can find a change of co-ordinates $y \mapsto Y$ under which β becomes $\pm y^b$. Note here that x, X generate the same ideal in $\mathcal{E}_1$, and likewise y, Y generate the same ideal in $\mathcal{E}_1$. Now we consider the possibilities regarding α, β.

<u>α, β are both zero</u> This case yields the germ $(xy, 0)$, which is not of finite $\mathcal{K}$-codimension, so can be discarded.

<u>Just one of α, β is zero</u> Suppose $\alpha \neq 0$. (A similar argument will apply in the case $\beta \neq 0$.) A change of co-ordinates $x \mapsto X$, $y \mapsto y$ brings the germ to the form $(Xy, \pm x^a)$, and since Xy, xy generate the same ideal this is $\mathcal{K}$-equivalent by (2.1) to (xy, x^a). But (xy, x^a) is also not of finite $\mathcal{K}$-codimension, so this case can likewise be discarded.

<u>α, β are both non-zero</u> A change of co-ordinates $x \mapsto X$, $y \mapsto Y$ brings the germ to the form $(XY, \pm x^a \pm y^b)$ and since xy, XY generate the same ideal this is $\mathcal{K}$-equivalent by (2.1) to $(xy, \pm x^a \pm y^b)$. Note at this point that we are at liberty to suppose $a \leq b$; and of course we can multiply either component by -1. Also, if a is odd we can change the sign in front of x^a by a change of co-ordinates $x \mapsto -x$, $y \mapsto y$: and similarly if b is odd we can change the sign in front of y^b. Thus in every case we obtain a germ of type $I_{a,b}$ save when the signs are different, and a, b are both even, and in that case we obtain a germ of type $II_{a,b}$.

<u>The $x^2 + y^2$ Case</u> Following the same initial reasoning used in the xy case we see that f is $\mathcal{K}$-equivalent to a germ $\left(x^2 + y^2, \zeta(x,y)\right)$ where ζ is a polynomial with no terms of degree ≤ 2. By (2.1) we do not change the $\mathcal{K}$-equivalence class by subtracting from ζ a multiple of $x^2 + y^2$ in $\mathcal{E}_2$: in particular we can suppose ζ has no terms with factor y^2, i.e. f has the form $\left(x^2 + y^2, \alpha(x) + y\beta(x)\right)$ with α, β polynomials. Note that

the second component cannot be identically zero, as $(x^2 + y^2, 0)$ has infinite $\mathcal{K}$-codimension. Denote by $a \geq 3$ the order of $\alpha(x) + y\beta(x)$, so the lowest order term is $px^a + qyx^{a-1}$, say, with at least one of $p, q \neq 0$. I claim we can suppose $p \neq 0$, $q = 0$. To this end consider a linear change of coordinates

$$\begin{cases} X = x \cos\theta + y \sin\theta \\ Y = -x \sin\theta + y \cos\theta \end{cases}$$

Notice that $x^2 + y^2 = X^2 + Y^2$. Using the fact that $Y^{2j} \equiv (-1)^j X^{2j}$ modulo $X^2 + Y^2$, a straightforward computation yields

$$px^a + qyx^{a-1} \equiv PX^a + QYX^{a-1} \quad \text{modulo } X^2 + Y^2$$

where

$$\begin{cases} P = p \cos a\theta - q \sin a\theta \\ Q = p \sin a\theta + q \cos a\theta \end{cases}$$

and the claim follows an observing that we can choose θ is such a way that $P \neq 0$, $Q = 0$. It follows that our germ is $\mathcal{K}$-equivalent to $(x^2 + y^2, x^a)$, which is the desired normal form of type IV_a. □

Of course one could go further, and take up the next case of germs of type $\Sigma^{2,1}$ and finite $\mathcal{K}$-codimension. At the time of writing no complete list is available, but one could certainly work with increasing codimension and gradually generate a list. However the reader can probably see for himself by now that such computations will become increasingly complicated and uninteresting. And that is as far as we shall pursue the problem of listing germs under the relation of $\mathcal{K}$-equivalence. The next step in our programme is to spell out just how all this enables us to list some of the simplest types of stable germs under A-equivalence.

§7. Some Examples of Classifying Stable Germs

We are now in a position to write out explicit lists of stable germs $(\mathbb{R}^n, 0) \to (\mathbb{R}^p, 0)$ under certain restrictions on the dimensions and the Boardman symbol. Let us start with the case $n \leq p$. The simplest possible situation is that of a non-singular germ. Such a germ is necessarily of type Σ^0, automatically stable, and has normal form $(x_1, \ldots, x_n, 0, \ldots, 0)$ by (I.1.4). We can reasonably expect the next simplest case to be stable germs of type Σ^1. Here we have a complete result, due to B. Morin.

(7.1) Let $n \leq p$, and let $F : (\mathbb{R}^n, 0) \to (\mathbb{R}^p, 0)$ be a stable germ of type Σ^1. Then F is necessarily of type $\Sigma^{1,\ldots,1,0}$ (with k repetitions) for some integer k with $1 \leq k \leq n/q$ where $q = p - n + 1$. And in that case F is A-equivalent to the germ $G : (\mathbb{R}^n, 0) \to (\mathbb{R}^p, 0)$ with components

$$\begin{cases} G_i = u_i & (1 \leq i \leq n-1) \\ G_{n+i} = \sum_{j=1}^{k} u_{ik+j} x^j & (0 \leq i \leq q-2) \\ G_p = \sum_{j=1}^{k-1} u_{(q-1)k+j} x^j + x^{k+1} \end{cases}$$

where we write $u_1, \ldots, u_{n-1}, x$ for the standard co-ordinates on $\mathbb{R}^n$.

Proof   The initial statement follows immediately from (5.4) and (6.4). The theory of §4 tells us that F must be A-equivalent to an $(n-1)$-parameter unfolding of a germ $f : (\mathbb{R}, 0) \to (\mathbb{R}^q, 0)$ of finite $\mathcal{K}$-codimension, and also of type $\Sigma^{1,\ldots,1,0}$ by (5.8). In view of (6.5) we know that f is $\mathcal{K}$-equivalent to $(0, \ldots, 0, x^{k+1})$, of $\mathcal{K}$-codimension $qk + q - 1$ by

Example 7 in §2. The theory now tells us that F is A-equivalent to the stable germ associated to a $\mathscr{K}$-versal deformation of the germ $(0, \ldots, 0, x^{k+1})$. The deformation in question was computed in Example 2 of §3, and the germ written out above is clearly the associated stable germ. Finally, the theory tells us that we need only consider those k for which the $\mathscr{K}$-codimension $qk + q - 1$ is $\leq p$, i.e. for which $k \leq n/q$. □

It is probably worthwhile isolating the <u>equidimensional case</u> of this result, i.e. the case $n = p$.

(7.2)   <u>Let</u> $F : (\mathbb{R}^n, 0) \to (\mathbb{R}^n, 0)$ <u>be a stable germ of type</u> Σ^1. F <u>must be of type</u> $\Sigma^{1,\ldots,1,0}$ <u>(with</u> k <u>repetitions) for some integer</u> k <u>with</u> $1 \leq k \leq n$, <u>and in that case is A-equivalent to the germ</u> $G : (\mathbb{R}^n, 0) \to (\mathbb{R}^n, 0)$ <u>with components</u>

$$\begin{cases} G_i = u_i & (1 \leq i \leq n - 1) \\ G_n = \sum_{j=1}^{k-1} u_j x^j + x^{k+1}. \end{cases}$$

Of course, this is the germ which we have previously dubbed the "generalized" Whitney mapping, and which in the special case $n = 2$, $k = 2$ yields the Whitney cusp mapping of the plane.

Let us now reverse the emphasis by taking up the case of stable germs $(\mathbb{R}^n, 0) \to (\mathbb{R}^p, 0)$ with $n \geq p$. Again, the simplest possible situation is that of a non-singular germ. Such a germ is necessarily of type Σ^{n-p}, automatically stable, and has normal form $(x_1, \ldots, x_p)$ by (I.1.3). We can reasonably expect the next simplest case to be that of stable germs of type Σ^{n-p+1}. The starting point here, as always, is that the theory of §4 tells us that such a germ is A-equivalent to a $(p-1)$-parameter unfolding of a germ $f : (\mathbb{R}^m, 0) \to (\mathbb{R}, 0)$ also of type Σ^{n-p+1}, where $m = n - p + 1$, and of finite $\mathcal{K}$-codimension. Our experience in Chapter IV tells us that the degree of complication in f depends at root on its corank c. Notice that the corank c, and the second order Boardman symbol, both depend solely on the 2-jet of f. It is therefore a good guess that the two invariants will be closely connected. Indeed that is the case.

(7.3)   <u>Let $f : (\mathbb{R}^m, 0) \to (\mathbb{R}, 0)$ be a germ of type Σ^m (i.e. singular). Then f has type $\Sigma^{m,c}$ if and only if f has corank c.</u>

Proof   We start with the condition for f to be of type $\Sigma^{m,c}$. The ideal $\Delta^m I_f$ is generated by f and the partial derivatives $\frac{\partial f}{\partial x_1}, \ldots, \frac{\partial f}{\partial x_m}$, and is to be critical. The ideal $\Delta^s \Delta^m I_f$ is generated by the same list, together with the minors of order $(m - s + 1)$ of their Jacobian

$$\begin{pmatrix} \frac{\partial f}{\partial x_1} & \cdots & \frac{\partial f}{\partial x_m} \\ \frac{\partial^2 f}{\partial x_1^2} & \cdots & \frac{\partial^2 f}{\partial x_1 \partial x_m} \\ \vdots & & \vdots \\ \frac{\partial^2 f}{\partial x_m \partial x_1} & \cdots & \frac{\partial^2 f}{\partial x_m^2} \end{pmatrix}$$

The condition for the ideal $\Delta^s\Delta^m I_f$ to be proper is that all its generators should have zero constant term, i.e. that all the minors of order $(m - s + 1)$ should vanish at 0. In this matrix we can disregard the first row, since all the $\frac{\partial f}{\partial x_j}$ vanish at 0, so are left precisely with the Hessian matrix of f. Thus the condition for the ideal $\Delta^s\Delta^m I_f$ to be proper is that the Hessian should have corank $\geq$ s, and the condition for $\Delta^s\Delta^m I_f$ to be critical is that the Hessian should have corank exactly s. Taking s = c we see that f has type $\Sigma^{m,c}$ if and only if f has corank c. □

The simplest situation is when f has corank 0. This yields another result of B. Morin.

(7.4)   Let $n \geq p$, and let $F : (\mathbb{R}^n, 0) \to (\mathbb{R}^p, 0)$ be a stable germ of type $\Sigma^{n-p+1,0}$: then F is A-equivalent to a germ $G : (\mathbb{R}^n, 0) \to (\mathbb{R}^p, 0)$ given by

$$\begin{cases} G_i = u_i & (1 \leq i \leq p-1) \\ G_p = \pm x_p^2 \pm \cdots \pm x_{n-1}^2 \pm x_n^2. \end{cases}$$

Proof   As was pointed out above, F is certainly A-equivalent to a $(p - 1)$-parameter unfolding of a germ $f : (\mathbb{R}^m, 0) \to (\mathbb{R}, 0)$ of type $\Sigma^{m,0}$ and finite $\mathcal{K}$-codimension, where $m = n - p + 1$. By (7.3) f has corank 0, so by the Morse Lemma is $\mathcal{R}$-equivalent, hence $\mathcal{K}$-equivalent, to a germ $\pm x_p^2 \pm \cdots \pm x_n^2$. The theory of §4 now tells us that F is A-equivalent to the stable germ associated to a p-parameter $\mathcal{K}$-versal deformation of this germ. It is a trivial computation to verify that the germ written out above is the stable germ in question.

The next simplest case is when f has corank 1, giving rise to the following result, also due to B. Morin.

(7.5) Let $n \geq p$, and let $F : (\mathbb{R}^n, 0) \to (\mathbb{R}^p, 0)$ be a stable germ of type $\Sigma^{n-p+1,1}$. Then F is necessarily of type $\Sigma^{n-p+1,1,\ldots,1,0}$ (with k repetitions) for some integer k with $1 \leq k \leq q$: and in that case F is A-equivalent to a germ $G : (\mathbb{R}^n, 0) \to (\mathbb{R}^p, 0)$ given by

$$\begin{cases} G_i = u_i \\ G_p = \pm x_p^2 \pm \ldots \pm x_{n-1}^2 \pm x_n^{k+2} + \sum_{j=1}^{k} u_j x_n^j . \end{cases}$$

Proof The initial statement follows immediately from (5.4) and (6.4). The rest of the proof follows exactly the same lines as that of (7.4), save that this time f is $\mathcal{K}$-equivalent to $\pm x_p^2 \pm \ldots \pm x_{n-1}^2 \pm x_n^{k+2}$ by (6.3). The computation of the deformation is very straightforward, and can safely be left to the reader. □

Of course, one could push these techniques further, by systematically classifying germs of functions with increasing $\mathcal{K}$-dimension. However the point of the last two results has already been made in that we have indicated the simplest complete results which can be obtained. Note incidentally that in the equidimensional case $n = p$ the last two results yield (7.2). Let us pursue the equidimensional case further. Having dealt with germs of type Σ^1, the next case to study is that of Σ^2 germs. In view of the discussion of §6 it is clear that we can only expect a complete result in the case of $\Sigma^{2,0}$ germs. The result is due to J. Mather.

(7.6) **Let** $F : (\mathbb{R}^n, 0) \to (\mathbb{R}^n, 0)$ **be a stable germ of type** $\Sigma^{2,0}$: **then** F **is A-equivalent to one of the following germs** $G : (\mathbb{R}^n, 0) \to (\mathbb{R}^n, 0)$.

Types $I_{a,b}$ and $II_{a,b}$
$$\begin{cases} G_i = u_i & (1 \le i \le a-1) \\ G_{a-1+j} = v_j & (1 \le j \le b-1) \\ G_{a+b-1} = xy \\ G_{a+b} = x^a \pm y^b + \sum_{i=1}^{a-1} u_i x^i + \sum_{j=1}^{b-1} v_j y^j \end{cases}$$

Type IV_a
$$\begin{cases} G_i = u_i & (1 \le i \le a-1) \\ G_{a-1+j} = v_j & (1 \le j \le a-1) \\ G_{2a-1} = x^2 + y^2 \\ G_{2a} = x^a + \sum_{i=1}^{a-1} u_i x^i + \sum_{j=1}^{a-1} v_j x^{j-1} y \end{cases}$$

Proof By the theory of §4 the germ F is A-equivalent to an $(n-2)$-parameter unfolding of a germ $f : (\mathbb{R}^2, 0) \to (\mathbb{R}^2, 0)$ of type $\Sigma^{2,0}$ and finite $\mathscr{K}$-codimension. By (6.7) this germ is $\mathscr{K}$-equivalent to one of the types $I_{a,b}$, $II_{a,b}$, IV_a. And then F will be A-equivalent to the stable germ associated to a $\overline{\mathscr{K}}$-versal deformation of these germs. The deformations were computed in Examples 3, 4 of §3, and the corresponding stable germs are evidently the germs written out above. □

With these results we have gone as far as is possible in a book of this nature, and have achieved our objective of indicating how one goes about classifying singular points of smooth mappings. We shall however pursue the matter a little further by looking briefly at certain mappings whose singular points are necessarily stable.

§8. Singular Points of Stable Mappings

We have so far discussed only stable <u>germs</u> of smooth mappings $f : \mathbb{R}^n \to \mathbb{R}^p$. But the underlying ideas can be introduced equally well for mappings themselves. By analogy with germs there is a natural notion of "equivalence" for smooth mappings; namely, we call two smooth mappings $f_1, f_2 : \mathbb{R}^n \to \mathbb{R}^p$ equivalent when there exist diffeomorphisms g, h for which the following diagram commutes

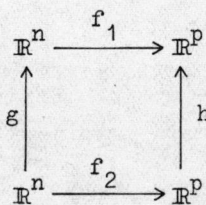

Now g is an element of the group $\text{Diff}(\mathbb{R}^n)$ of all diffeomorphisms of $\mathbb{R}^n$, and h is an element of the group $\text{Diff}(\mathbb{R}^p)$ of all diffeomorphisms of $\mathbb{R}^p$. So the pair (g, h) is an element of the product group $\text{Diff}(\mathbb{R}^n) \times \text{Diff}(\mathbb{R}^p)$ and this group acts on $C^\infty(\mathbb{R}^n, \mathbb{R}^p)$ if we define $(g, h).f$ to be $h \circ f \circ g^{-1}$. Of course this action lies outside the framework discussed in Chapter III because neither the group $\text{Diff}(\mathbb{R}^n) \times \text{Diff}(\mathbb{R}^p)$ nor the set $C^\infty(\mathbb{R}^n, \mathbb{R}^p)$ are in any way finite-dimensional. But for all that there is no harm in proceeding by analogy, just as we did with germs. Recall that in Chapter III we introduced the equivalent notions of "stability" and "infinitesimal stability", the former being geometric in nature, and the latter algebraic. When dealing with germs we interpreted stability in terms of unfoldings, but for mappings it will be easier if we stick to the geometric idea. We should therefore call $f : \mathbb{R}^n \to \mathbb{R}^p$ "stable" when all "sufficiently close" $g : \mathbb{R}^n \to \mathbb{R}^p$ are equivalent to f. On this basis we introduce the following formal definition.

A smooth mapping $f : \mathbb{R}^n \to \mathbb{R}^p$ is <u>stable</u> when there exists a real number $\epsilon > 0$ such that every smooth mapping $g : \mathbb{R}^n \to \mathbb{R}^p$ is the ϵ-neighbourhood of f is equivalent to f. Our first result gives us some idea of just how nice stable mappings are.

(8.1) <u>Any stable mapping $f : \mathbb{R}^n \to \mathbb{R}^p$ is generic in the sense of Boardman.</u>

<u>Proof</u> Let $k \geq 1$ be an integer. We have to show that $j^k f$ is transverse to all the Boardman submanifolds $\Sigma^{i_1,\ldots,i_k}$. To this end recall from (5.6) that the set of all smooth mappings $\mathbb{R}^n \to \mathbb{R}^p$ with this property is dense in $C^\infty(\mathbb{R}^n, \mathbb{R}^p)$: thus, given any $\epsilon > 0$ we can find a $g : \mathbb{R}^n \to \mathbb{R}^p$ in the ϵ-neighbourhood of f for which $j^k g$ is transverse to all the $\Sigma^{i_1,\ldots,i_k}$. And since f is stable we can choose ϵ so small that every map in the ϵ-neighbourhood of f is equivalent to f. It remains only to observe that if f is equivalent to g, and g satisfies a transversality condition, then f does as well - which fact we leave as an exercise for the reader. □

(8.2) <u>Let $f : \mathbb{R}^n \to \mathbb{R}^p$ be stable: then the germ of f at any point is stable.</u>

We shall omit the proof of this result. If we now combine these results with the classifications obtained in the previous paragraphs we obtain some aesthetically very pleasing theorems describing completely all possible singular points of a stable mapping $f : \mathbb{R}^n \to \mathbb{R}^p$ for certain values of n, p.

Let us start by looking at the equidimensional case $n = p$. Consider a stable mapping $f : \mathbb{R}^n \to \mathbb{R}^n$. First of all, f must be transverse to the first-order Boardman submanifolds Σ^i - which by (5.1) in Chapter II have codimension i^2. In particular, if $n \leq 3$ the first-order singularity sets $\Sigma^i f$ — which have the same codimension — must be empty for $i \geq 2$, and the only possible singular points are those of type Σ^1. In fact we can be more explicit. Recall that in the equidimensional case the singularity set $\Sigma^{1,\ldots,1,0} f$ (with k repetitions) has codimension k, by Example 8 of §5: thus the only singularity sets $\Sigma^{1,\ldots,1,0} f$ which can be non-void are those with $k \leq n$. If we take $n = 1$ we obtain immediately from (7.2)

(8.3)   <u>Let $f : \mathbb{R} \to \mathbb{R}$ be a stable mapping: its germ at any point is equivalent to one of the following</u>

$$\Sigma^0 \;:\; y_1 = x_1 \qquad \text{(regular)}$$

$$\Sigma^{1,0} \;:\; y_1 = x_1^2 \qquad \text{(simple minimum)}.$$

Taking the next case $n = 2$ we recover a famous result of H. Whitney describing the possible singularities of stable mappings from the plane to itself.

(8.4)   <u>Let $f : \mathbb{R}^2 \to \mathbb{R}^2$ be a stable mapping: its germ at any point is equivalent to one of the following</u>

$$\Sigma^0 \quad \begin{cases} y_1 = x_1 \\ y_2 = x_2 \end{cases} \qquad \text{(regular)}$$

$$\Sigma^{1,0} \quad \begin{cases} y_1 = x_1 \\ y_2 = x_2^2 \end{cases} \qquad \text{(fold)}$$

$$\Sigma^{1,1,0} \begin{cases} y_1 = x_1 \\ y_2 = x_2^3 + x_1 x_2 \end{cases} \qquad \text{(cusp)}.$$

In order to obtain geometric insight into these germs it is best to think of them respectively as composites

$$(x_1, x_2) \longmapsto (x_1, x_2, x_2) \xrightarrow{\text{proj}} (x_1, x_2)$$

$$(x_1, x_2) \longmapsto (x_1, x_2, x_2^2) \xrightarrow{\text{proj}} (x_1, x_2^2)$$

$$(x_1, x_2) \longmapsto (x_1, x_2, x_2^3 + x_1 x_2) \xrightarrow{\text{proj}} (x_1, x_2^3 + x_1 x_2)$$

where in each case the second mapping is (the restriction to the image of the first of) the projection $(x_1, x_2, x_3) \to (x_1, x_3)$. In this way we can visualize our maps as follows.

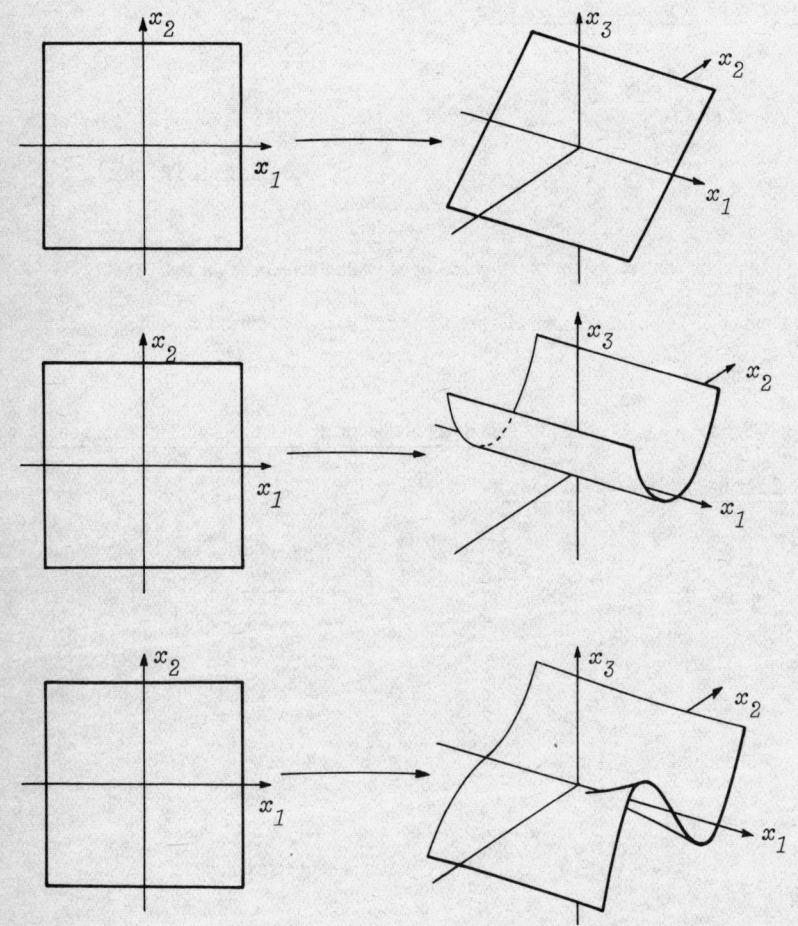

Take for instance the situation of Example 4 in §4 of Chapter II, namely the projection of a torus onto a plane. Locally, such a mapping is from the plane to itself, and generically such a mapping is stable - a fact which we shall have to ask our reader to accept - so we should only see singular points of the above three types. Indeed this is the case. Recall that the set of critical values looks like this.

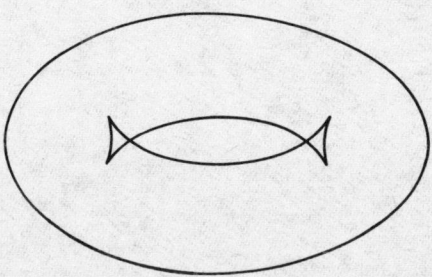

Here the curves which make up the picture are the images under the projection f of the singularity sets $\Sigma^{1,0} f$: at a point in such a set we simply see the torus folding over. But the four exceptional points are the images of the singularity sets $\Sigma^{1,1} f$: here we have the more complicated situation of two folds coming together to form a "pleat".

We can of course gain one more result by taking the final case $n = 3$.

(8.5) <u>Let $f : \mathbb{R}^3 \to \mathbb{R}^3$ be a stable mapping: its germ at any point is equivalent to one of the following.</u>

$$\Sigma^0 \begin{cases} y_1 = x_1 \\ y_2 = x_2 \\ y_3 = x_3 \end{cases} \qquad \text{(regular)}$$

$$\Sigma^{1,0} \begin{cases} y_1 = x_1 \\ y_2 = x_2 \\ y_3 = x_3^2 \end{cases} \quad \text{(fold)}$$

$$\Sigma^{1,1,0} \begin{cases} y_1 = x_1 \\ y_2 = x_2 \\ y_3 = x_3^3 + x_1 x_3 \end{cases} \quad \text{(cusp)}$$

$$\Sigma^{1,1,1,0} \begin{cases} y_1 = x_1 \\ y_2 = x_2 \\ y_3 = x_3^4 + x_1 x_3 + x_2 x_3^2 \end{cases} \quad \text{(dovetail)}$$

Of course, when we go up to the next dimension $n = 4$ we no longer obtain only Σ^1 points. As we have already pointed out $\Sigma^i f$ has codimension i^2 in the equidimensional case so when $n = 4$ we can certainly have Σ^2 points, but avoid Σ^i points for $i \geq 3$. Now $\Sigma^2 f$ splits into $\Sigma^{2,0} f$, $\Sigma^{2,1} f$, $\Sigma^{2,2} f$ with codimensions 4, 7, 10 respectively: thus only the first of these three sets can be non-void. Appealing to the classification of stable germs of type $\Sigma^{2,0}$ given in §7 we obtain the following:

(8.6) Let $f : \mathbb{R}^4 \to \mathbb{R}^4$ be a stable mapping: its germ at any point is equivalent to one of the following.

$$\Sigma^0 \begin{cases} y_i = x_i \quad (1 \leq i \leq 3) \\ y_4 = x_4 \end{cases}$$

$\Sigma^{1,0}$ $\begin{cases} y_i = x_i & (1 \leq i \leq 3) \\ y_4 = x_4^2 \end{cases}$

$\Sigma^{1,1,0}$ $\begin{cases} y_i = x_i & (1 \leq i \leq 3) \\ y_4 = x_4^3 + x_1 x_4 \end{cases}$

$\Sigma^{1,1,1,0}$ $\begin{cases} y_i = x_i & (1 \leq i \leq 3) \\ y_4 = x_4^4 + x_1 x_4 + x_2 x_4^2 \end{cases}$

$\Sigma^{1,1,1,1,0}$ $\begin{cases} y_i = x_i & (1 \leq i \leq 3) \\ y_4 = x_4^5 + x_1 x_4 + x_2 x_4^2 + x_3 x_4^3 \end{cases}$

$\Sigma^{2,0}$
$I_{2,2,2}$ $\begin{cases} y_1 = x_1 \\ y_2 = x_2 \\ y_3 = x_3 x_4 \\ y_4 = x_3^2 + x_4^2 + x_1 x_3 + x_2 x_4 \end{cases}$

$\Sigma^{2,0}$
$I_{2,2}$ $\begin{cases} y_1 = x_1 \\ y_2 = x_2 \\ y_3 = x_3 x_4 \\ y_4 = x_3^2 - x_4^2 + x_1 x_3 + x_2 x_4 \end{cases}$.

The reader is invited to continue the listing process for stable mappings $\mathbb{R}^5 \to \mathbb{R}^5$, $\mathbb{R}^6 \to \mathbb{R}^6$. That is as far as he will get by just dipping into the results of the previous sections. Stable mappings $\mathbb{R}^7 \to \mathbb{R}^7$ can have germs of type $\Sigma^{2,1}$, and these we have not listed: however, the enterprising reader will find that he can obtain a complete list in this case too, using a little common sense and the techniques already expounded.

Let us diverge from the equidimensional case to see what further gain can be derived from the results of the previous three sections. Take for instance the case $n \le p$. We ask if there are pairs of integers (n, p) with $n \le p$ having the property that a stable mapping $\mathbb{R}^n \to \mathbb{R}^p$ admits only singular points of type Σ^1. We can determine a useful range of such pairs (n, p) by a simple combinatoric exercise. We know that Σ^i has codimension $i(p - n + i)$. If our mapping is to admit singular points of type Σ^1 we must have $i(p - n + i) \le n$ in the case $i = 1$, i.e. $p < 2n$. But for $i \ge 2$ we do not wish our mapping to have singular points of type Σ^i, so need $i(p - n + i) > n$ for $i \ge 2$, i.e. $2p > 3n - 4$. We should certainly then be looking at pairs (n, p) in the range $3n - 4 < 2p < 4n$.

It is worthwhile analysing the situation a little further. Suppose we have a pair (n, p) in this range: we ask which singular points of type $\Sigma^{1,\ldots,1,0}$ can arise. Recall that the codimension of $\Sigma^{1,\ldots,1,0}$ (with k repetitions) is $(p - n + 1)k$. The condition $p < 2n$ ensures that for $k - 1$ $k - 1$ we have $(p - n + 1)k \le n$, so we are not excluding the possibility of having $\Sigma^{1,0}$ points. We shall certainly avoid singular points of type $\Sigma^{1,\ldots,1,0}$ (with two or more repetitions) if $(p - n + 1)k > n$ for $k \ge 2$, i.e. $2p \ge 3n - 1$, imposing only a very slight further restriction. A natural range to consider then is given by $3n - 1 \le 2p < 4n$, which is sometimes called the <u>metastable range</u>. Combining these arguments with the results of §7 we deduce the following result, which in the case $p = 2n - 1$ is due to H. Whitney.

(8.7)   Let $f : \mathbb{R}^n \to \mathbb{R}^p$ be a stable mapping with the pair (n, p) in the metastable range: then f only admits singular points of type $\Sigma^{1,0}$, and its germ at such a point is equivalent to the germ

$$\begin{cases} y_i = x_i & (1 \leq i \leq n - 1) \\ y_{n-1+i} = x_i x_n & (1 \leq i \leq p - n) \\ y_p = x_n^2 \end{cases}$$

The case $n = 1$ of this result is simply (8.3). But when $n = 2$ we get something new, namely that a stable mapping $f : \mathbb{R}^2 \to \mathbb{R}^3$ only admits singular points of type $\Sigma^{1,0}$ and that the germ at such a point is equivalent to the germ

$$\begin{cases} y_1 = x_1 \\ y_2 = x_1 x_2 \\ y_3 = x_2^2 \end{cases}$$

whose image is the so-called Whitney umbrella depicted below.

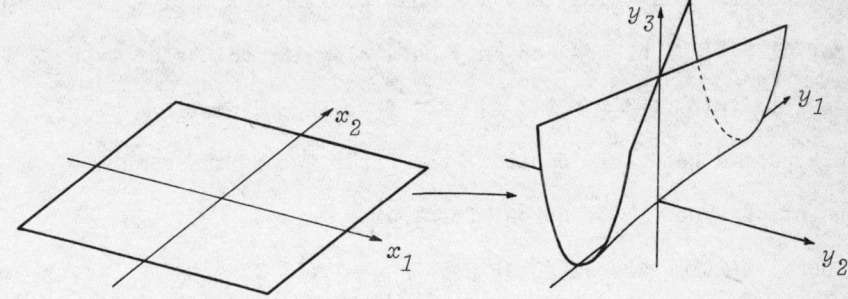

213

What about the case $n \geq p$? Here again one could construct ranges of pairs (n, p) for which a stable mapping $\mathbb{R}^n \to \mathbb{R}^p$ assumes only singular points of a very simple type. However the combinatorics are rather more complex, and it is probably more illuminating to take specific pairs (n, p) and write out a complete list of possibilities. One of the simplest cases is

(8.8) <u>Let $f : \mathbb{R}^3 \to \mathbb{R}^2$ be a stable mapping: its germ at any point is equivalent to one of the following.</u>

$$\Sigma^1 \quad \begin{cases} y_1 = x_1 \\ y_2 = x_2 \end{cases}$$

$$\Sigma^{2,0} \quad \begin{cases} y_1 = x_1 \\ y_2 = \pm x_2^2 \pm x_3^2 \end{cases}$$

$$\Sigma^{2,1,0} \quad \begin{cases} y_1 = x_1 \\ y_2 = \pm x_2^2 \pm x_3^3 + x_1 x_3 \end{cases}.$$

<u>Proof</u>   In this situation $\Sigma^i f$ has codimension $i(i-1)$ so must be empty for $i \geq 3$, i.e. we can only have singular points of type Σ^2. Now $\Sigma^2 f$ splits into $\Sigma^{2,0} f$, $\Sigma^{2,1} f$, $\Sigma^{2,2} f$ with codimensions 2, 3, 5 respectively, so we need only consider the first two of these sets. At a singular point of type $\Sigma^{2,0}$ normal forms are provided by (7.4). The only remaining possibilities are singular points of type $\Sigma^{2,1,\ldots,0}$ (with k repetitions): this has codimension $k + 2$, as an easy computation verifies, so we need only consider the cases $k = 0, 1$ with normal forms provided by (7.5).

$\square$

Appendix A The theorem of Sard

The key idea here is borrowed from measure theory, and is that of a "null set"; this will provide our starting point. A set $V \subseteq \mathbb{R}^n$ is called a null set in $\mathbb{R}^n$ when, given any real number $\epsilon > 0$, V is contained in a countable union of cubes the sum of whose volumes is $< \epsilon$. (A cube in $\mathbb{R}^n$ is a product of n open intervals, called its sides, and its volume is the product of their lengths.) The reader will readily check that in using this definition one can assume that each cube has all its sides of equal length. Note that when V is compact the countable union of cubes can be assumed finite. The first fact we need is

(A1) Any countable union V of null sets V_j in $\mathbb{R}^n$ is again a null set in $\mathbb{R}^n$.

Proof   Let $\epsilon > 0$. We can cover V_j by countably many cubes $C_{j,k}$ of total volume $< \epsilon/2^j$: hence we can cover V by the countably many cubes $C_{j,k}$ of total volume $< \sum_j \epsilon/2^j \leq \epsilon$.   □

It seems worthwhile spelling out a simple consequence of (A1) which we shall use several times in the proof of Sard's Theorem.

(A2)   Let $f : X \to Y$ be a mapping with $X \subseteq \mathbb{R}^n$, $Y \subseteq \mathbb{R}^p$. If each point $x \in X$ has a neighbourhood U_x in $\mathbb{R}^n$ for which $f(U_x \cap X)$ is a null set in $\mathbb{R}^p$ then $f(X)$ is a null set in $\mathbb{R}^p$.

<u>Proof</u>   From the cover $(U_x)_{x \in X}$ for X one can extract a countable subcover, so $f(X)$ is a countable union of null sets in $\mathbb{R}^p$, hence a null set in $\mathbb{R}^p$, by (A1).

The next fact that we need to know is that the property of being a null set is invariant under smooth mappings, in the following precise sense.

(A3)   Let $f : N \to P$ <u>be a smooth mapping with</u> N, P <u>open sets in</u> $\mathbb{R}^n$: <u>if</u> $V \subseteq N$ <u>is a null set in</u> $\mathbb{R}^n$ <u>then likewise</u> $f(V) \subseteq P$ <u>is a null set in</u> $\mathbb{R}^n$.

<u>Proof</u>   By (A2) it will suffice to show that $f(C \cap V)$ is a null set in $\mathbb{R}^n$ for each cube C in $\mathbb{R}^n$ whose closure is contained in N. It follows from the Mean Value Theorem that there exists a real number $K > 0$ such that for any x, y in C

$$|f(x) - f(y)| \leq K|x - y| .$$

Choose $\epsilon > 0$. Since $C \cap V$ is a null set in $\mathbb{R}^n$ it can be covered by countably many cubes $C_i \subseteq C$, each C_i having all its sides of equal length c_i, and of total volume $\sum_i c_i^n < \epsilon$. It follows that each $f(C_i)$ is contained in a cube having all its sides of equal length $(Kn^{\frac{1}{2}})c_i$, and hence that $f(C \cap V)$ is covered by a countable family of cubes of total volume $\sum_i (Kn^{\frac{1}{2}})^n c_i^n < (Kn^{\frac{1}{2}})^n \epsilon$. It follows that $f(C \cap V)$ is a null set in $\mathbb{R}^n$. $\square$

The next proposition is a very special case of Fubini's Theorem. To state it we need a little notation. Suppose $V \subseteq \mathbb{R}^n = \mathbb{R} \times \mathbb{R}^{n-1}$. For $s \in \mathbb{R}$ define

$$V_s = \{(x_1, \ldots, x_{n-1}) \in \mathbb{R}^{n-1} : (s, x_1, \ldots, x_{n-1}) \in V\}.$$

(A4) <u>Let $V \subseteq \mathbb{R}^n$ be countable union of compact sets such that for all $s \in \mathbb{R}$ the set $V_s \subseteq \mathbb{R}^{n-1}$ is a null set in $\mathbb{R}^{n-1}$: then V is a null set in $\mathbb{R}^n$.</u>

<u>Step 1</u>   Suppose first that V is compact, so there exists a closed interval I for which $V \subseteq I \times \mathbb{R}^{n-1}$. Now let $s \in I$ and let $\epsilon > 0$. By hypothesis V_s can be covered by finitely many cubes $C_{s,j}$ in $\mathbb{R}^{n-1}$ of total volume $< \epsilon$: indeed it follows from the compactness of V that there is an open interval I_s, with mid-point s, having the property that the same cubes cover all the V_t for t in I_s. From this cover of I by open intervals I_s we can extract a finite subcover $I_{s_1}, \ldots, I_{s_p}$: one can certainly suppose that this subcover is minimal, i.e. that if we delete any one of these intervals the remaining intervals will fail to cover I, and hence that the sum of their lengths is not more than twice the length L of I. Thus we have finitely many cubes $I_{s_i} \times C_{s_i,j}$ covering V, and their total volume is $\leq 2\epsilon L$. It follows that V is a null set in $\mathbb{R}^n$.

<u>Step 2</u>   Now suppose V is a countable union of compact sets. It follows immediately from Step 1 and (A1) that V is a null set in $\mathbb{R}^n$.   □

So much for technical preliminaries. The next step is to extend the basic notion of a null set to subsets of smooth manifolds. Let $N^n \subseteq \mathbb{R}^s$ be a smooth manifold. A set $V \subseteq N$ is called a <u>null set in N</u> when for every chart $\phi : A \cap N \to B$ one has $\phi(A \cap V)$ a null set in $\mathbb{R}^n$: of course A, B are open sets in $\mathbb{R}^s$, $\mathbb{R}^n$ respectively. Sard's Theorem is the following proposition.

(A5)   <u>Let $f : N \to P$ be a smooth mapping with N^n, P^p smooth manifolds. The set of critical values of f is a null set in P.</u>

Proof   One proceeds by induction on n. The case $n = 0$ is trivial. (It is a convention that $\mathbb{R}^0$ comprises a single point.) We suppose therefore that (A5) holds for $(n - 1)$, and will deduce that it holds for n. It follows immediately from the definitions that we can suppose N, P to be open sets in $\mathbb{R}^n$, $\mathbb{R}^p$.

Remark   It is probably worthwhile pointing out that the case $n < p$ is relatively trivial. Consider indeed the smooth mapping $N \times \mathbb{R}^{p-n} \to P$ given by $(x, y) \to f(x)$. One easily checks that $N \times 0$ is a null set in $\mathbb{R}^p$, so that its image under this mapping, namely $f(N)$, is a null set in $\mathbb{R}^p$ by (A3). It remains only to observe that when $n < p$ the set of critical values of f is precisely $f(N)$. Of course when $n \geq p$ this reasoning breaks down and one needs a more subtle argument: this shows that we may as well assume $n \geq p$, so the critical points are precisely the singular points.

Proof (continued)   Write Σ for the singular set of f. And write Σ_k for the set of points $x \in N$ where all the partial derivatives of order $\leq k$ are zero. Thus

$$\Sigma \supseteq \Sigma_1 \supseteq \Sigma_2 \supseteq \dots .$$

We have to show that $f(\Sigma)$ is a null set in $\mathbb{R}^p$. In view of (A1) it will suffice to establish the following steps.

  Step 1 : $f(\Sigma - \Sigma_1)$ is a null set in $\mathbb{R}^p$.
  Step 2 : $f(\Sigma_k - \Sigma_{k+1})$ is a null set in $\mathbb{R}^p$, for all k.
  Step 3 : $f(\Sigma_k)$ is a null set in $\mathbb{R}^p$, for some k.

Step 1   Note first that we can assume $p \geq 2$ since when $p = 1$ we have $\Sigma = \Sigma_1$, and there is nothing to prove. By (A2) it suffices to show that each point x in $\Sigma - \Sigma_1$ has a neighbourhood U in N for which the image of $\Sigma - \Sigma_1$ under the restriction $f|U$ is a null set in $\mathbb{R}^p$. And using (A3) it is clear that we can replace this restriction by any equivalent mapping, which we shall also denote f. In particular we can assume $x = 0$, $f(x) = 0$: and since the rank of f at x is ≥ 1 and $< p$ we can suppose (by choosing linearly adapted local co-ordinates) that $f(s, x) = \left(s, f_s(x)\right)$ where $s \in \mathbb{R}$, and each f_s is a smooth mapping from an open set in $\mathbb{R}^{n-1}$ to an open set in $\mathbb{R}^{p-1}$.

We have to show that the set V of critical values of f is a null set in $\mathbb{R}^p$. Notice that the set of critical points of f is a closed set, hence a countable union of compact sets: it follows that V, the image of the set of critical points under f, is likewise a countable union of compact sets. With the notation of (A4) observe that V_s is precisely the set of critical values of f_s, which is a null set in $\mathbb{R}^{p-1}$, by the induction hypothesis for the theorem. It follows from (A4) that V is a null set.

<u>Step 2</u>   By (A2) it suffices to show that each point x in $\Sigma_1 - \Sigma_{k+1}$ has a neighbourhood U in N for which the image of $U \cap (\Sigma_k - \Sigma_{k+1})$ under f is a null set in $\mathbb{R}^p$. At x some $(k + 1)^{th}$ order partial derivative of f is $\neq 0$, say $\frac{\partial g}{\partial x_j}$ with g a kth order partial derivative of f, so there certainly exists a neighbourhood U of x whose intersection G with the set defined by $g = 0$ is a smooth manifold of dimension $(n - 1)$. Note that $\Sigma_k \subseteq G$: in particular $x \in G$. Clearly if U is small enough $U \cap (\Sigma_k - \Sigma_{k+1}) = U \cap \Sigma_k$ is contained in the set of critical points of $f|G$, so its image under f is a null set in $\mathbb{R}^p$, by the induction hypothesis of the theorem.

<u>Step 3</u>   Let k be an integer $> \frac{n}{p} - 1$. We shall show that $f(\Sigma_k)$ is a null set in $\mathbb{R}^p$. By (A2) it is enough to show that $f(C \cap \Sigma_k)$ is a null

set in $\mathbb{R}^p$ for any cube C in $\mathbb{R}^n$, of side c say, whose closure is contained in N. As a consequence of Taylor's Theorem there exists a real number $\kappa > 0$ for which

$$|f(x) - f(y)| \leq \kappa |x - y|^{k+1}$$

for all x, y in $C \cap \Sigma_k$. Introduce an arbitrary integer $r \geq 1$, and subdivide C into $R = r^n$ equal cubes $C_1, \ldots, C_R$ of side c/r. By the above inequality $f(C \cap \Sigma_k)$ is contained in a cube in $\mathbb{R}^p$ of side $2\kappa(c/r)^{k+1}$, and hence $f(C \cap \Sigma_k)$ is contained in the union of r^n cubes in $\mathbb{R}^p$ of total volume

$$r^n \cdot (2\kappa)^p (c/r)^{p(k+1)} = \text{constant} \times r^{(n-pk-p)}.$$

Note that $n - pk - p < 0$ as $k > \frac{n}{p} - 1$, so that this last expression $\to 0$ as $r \to \infty$. It follows immediately that $f(C \cap \Sigma_k)$ is a null set in $\mathbb{R}^p$. $\square$

In order to establish the consequence of Sard's Theorem used in Chapter II to prove the Basic Transversality Lemma we need

(A6)   Let P be a smooth manifold and $V \subseteq P$ a null set: then the complement of V is dense in P.

Proof   It follows immediately from the definitions that we can assume $P = \mathbb{R}^p$. Suppose (A6) were false, so we could find a cube C whose closure $\overline{C}$ is a subset of V. One derives a contradiction as follows. Since V is a null set one can, given $\epsilon > 0$, find a countable family of cubes (C_i) with $V \subseteq \cup C_i$ and with $\sum \text{vol}(C_i) < \epsilon$. As $\overline{C}$ is compact we can extract from the open cover (C_i) a finite subcover $C_1, \ldots, C_t$, say; then

$$\text{vol}(C) \leq \sum_{i=1}^{t} \text{vol}(C_i) \leq \sum \text{vol}(C_i) < \epsilon$$

and ϵ being arbitrary one deduces that vol $(C) = 0$, which is the desired contradiction. □

Note that we assume in the above proof that if $C \subseteq \bigcup_{i=1}^{t} C_i$ then automatically vol $(C) \leq \sum_{i=1}^{t}$ vol (C_i): we leave this fact as an exercise for the reader. Finally, we deduce

(A7) **Let $f_i : N_i \to P$ be a countable family of smooth mappings with the N_i, P smooth manifolds: the set of common regular values of the f_i is dense in P.**

Proof The set C_i of critical values of f_i is a null set in P, by Sard's Theorem, so the union $C = \bigcup C_i$ is a null set by (A1). The set of common regular values is the complement of C, so dense in P by (A6). □

Appendix B   Semialgebraic group actions

Let $\Phi : G \times M \to M$ be a smooth action of a Lie group G on a smooth manifold M. The objective of this appendix is to show that if the action is "semialgebraic", in a sense to be made precise below, then automatically all the orbits will be smooth submanifolds of M. The virtue of this fact is that most of the examples which arise in this area of mathematics turn out to be "semialgebraic", so that automatically the theory of Chapter III applies to them. At the time of writing the only reference for the theory of semialgebraic sets and mappings is the set of research notes by S. Lojasiewicz entitled "Ensembles Semi-Analytiques": and these, unfortunately, are not easily available. Until a more accessible account of the theory appears these notes must remain the sole reference for the two basic results we need, namely (B2) and (B3) below.

The ideas involved have their genesis in real algebraic geometry. Recall that a set $A \subseteq \mathbb{R}^n$ is <u>algebraic</u> when it can be obtained by finitely many applications of the operation of intersection, starting from sets of the form $\{x \in \mathbb{R}^n : f(x) = 0\}$ with f a polynomial function on $\mathbb{R}^n$. For instance, a little linear algebra shows that any linear subspace of $\mathbb{R}^n$ is algebraic. However, there are areas of mathematics where it is profitable to introduce a wider class of sets, closed under as many set theoretic and topological operations as is possible. One such class is obtained by calling $A \subseteq \mathbb{R}^n$ <u>semialgebraic</u> when it can be obtained by finitely many applications of the operations of intersection, union and set difference starting from sets of the form $\{x \in \mathbb{R}^n : f(x) > 0\}$ with f a polynomial function on $\mathbb{R}^n$. The reader will readily check that an algebraic set is automatically semialgebraic. A good example is provided by the general linear group $GL(s)$. This is a subset of the linear space $M(s)$ of all real $s \times s$ matrices, identified in an obvious way with a Euclidean space: the singular matrices form an algebraic subset of $M(s)$, given by the vanishing of the determinant, so that the complement $GL(s)$ is semialgebraic.

The idea can be extended to mappings in an obvious way. A mapping $f : A \to \mathbb{R}^p$ with $A \subseteq \mathbb{R}^n$ is <u>semialgebraic</u> when graph f is semialgebraic in $\mathbb{R}^n \times \mathbb{R}^p$. Linear projections provide simple examples of semialgebraic mappings, since their graphs are linear spaces. A wider class of examples arises from considering <u>rational</u> mappings $f : A \to \mathbb{R}^p$, i.e. those for which each component $f_i = \phi_i/\psi_i$ where ϕ_i, ψ_i are polynomial functions on $\mathbb{R}^n$, with ψ_i nowhere zero on A. It seems worthwhile spelling out the following proposition.

(B1)   <u>If $A \subseteq \mathbb{R}^n$ is semialgebraic, and $f : A \to \mathbb{R}^p$ is rational, then f is semialgebraic.</u>

<u>Proof</u>   With the above notation define polynomial functions $\theta_1, \ldots, \theta_p$ on $\mathbb{R}^n \times \mathbb{R}^p$ by

$$\theta_i(x, y) = \psi_i(x) - y_i \phi_i(x)$$

where $x = (x_1, \ldots, x_n)$ is in $\mathbb{R}^n$, and $y = (y_1, \ldots, y_p)$ is in $\mathbb{R}^p$. The vanishing of θ_i defines an algebraic subset H_i of $\mathbb{R}^n \times \mathbb{R}^p$. The proposition follows on noting that

$$\text{graph } f = (A \times \mathbb{R}^p) \cap H_1 \cap \ldots \cap H_p$$

is necessarily semialgebraic.   □

One of the basic facts about semialgebraic mappings is the Tarski-Seidenberg theorem.

(B2)   Let $X \subseteq \mathbb{R}^n$ be semialgebraic, and let $f : A \to \mathbb{R}^p$ be semialgebraic with $A \subseteq \mathbb{R}^n$: then the image $f(X)$ is semialgebraic in $\mathbb{R}^p$.

It follows from the Tarski-Seidenberg theorem that the domain of any semialgebraic mapping is necessarily semialgebraic, being the image of the graph under a linear projection. And by the same argument, if a product $A \times B$ is semialgebraic then so too are the factors A, B. We are going to be interested primarily in group actions $\Phi : G \times M \to M$ which are semialgebraic in the sense just described. It follows from the preceding remarks that then automatically G, M have to be semialgebraic.

We need one more fact. First, a preliminary definition. Let $A \subseteq \mathbb{R}^n$. A point $x \in A$ is regular (of dimension d) when x has a neighbourhood U in $\mathbb{R}^n$ for which $U \cap A$ is a smooth submanifold of $\mathbb{R}^n$ (of dimension d).

(B3)   Let $A \subseteq \mathbb{R}^n$ be a non-void semialgebraic set: then A has at least one regular point.

Now we can put the bits together to obtain the main result.

(B4)   Let $\Phi : G \times M \to M$ be a smooth action of a Lie group G on a smooth manifold M. And suppose that the action is semialgebraic. Then all the orbits are smooth submanifolds of M.

Proof   Let $x_0 \in M$. The orbit $G.x_0$ through x_0 is the image under Φ of the semialgebraic set $G \times \{x_0\}$, hence semialgebraic by the Tarski-Seidenberg theorem. By (B3) the orbit has at least one regular point, of dimension d, say. But the homogeneity property for orbits implies that then every point on the orbit is regular, of dimension d, i.e. the orbit is a smooth submanifold of M of that dimension. □

By way of explicit illustration consider the class of geometric actions studied in some detail in Chapter III. We have

(B5)   The natural action of $GL(n) \times GL(p)$ on $H^d(n, p)$ given by $(g, h).f = h \circ f \circ g^{-1}$ is semialgebraic, and hence all the orbits are smooth manifolds.

Proof $GL(n)$, $GL(p)$ and $H^d(n, p)$ are all semialgebraic, so their product, i.e. the domain of the action, is likewise semialgebraic. In view of (B1) it will now suffice to show that the action is rational. For this it will be convenient to think of our linear mappings g, h as square matrices (g_{ij}), (h_{ij}) respectively. We have to show that each coefficient of each component of $h \circ f \circ g^{-1}$ is a rational function of the entries in g, h and the coefficients of components of f. Now the ith component of $h \circ f \circ g^{-1}$ is precisely $h_{i1}F_1 + \ldots + h_{ip}F_p$, where $F_1, \ldots, F_p$ denote the components of $F = f \circ g^{-1}$; it will therefore be sufficient to show that each coefficient of each component of F is a rational function of the entries in g, and the coefficients of components of f. Certainly the entries g^*_{ij} of g^{-1} are rational functions of the entries g_{ij} of g, with nowhere-zero denominator. And the claim follows since $F_1, \ldots, F_p$ are obtained respectively from the components $f_1, \ldots, f_p$ of f by substituting $g^*_{i1}x_1 + \ldots + g^*_{in}x_n$ for the variable x_i. □

Appendix C   Real algebras

By a <u>real algebra</u> is meant a real vector space V together with a bilinear mapping $V \times V \to V$, written $(x, y) \to x.y$ and called the <u>algebra product</u>. In this book only two examples of algebras are of interest.

<u>Example 1</u>   The real vector space $\mathcal{E}_n$ of all germs $f : (\mathbb{R}^n, 0) \to (\mathbb{R}, y)$ of smooth functions. This is endowed with the algebra product induced from that on the reals.

<u>Example 2</u>   The real vector space $\hat{\mathcal{E}}_n$ of all formal power series $\hat{f} = \sum f_\alpha x^\alpha$ in n real indeterminates $x_1, \ldots, x_n$. Here, given $x = (x_1, \ldots, x_n)$ and $\alpha = (\alpha_1, \ldots, \alpha_n)$ we write x^α as an abbreviation for the expression $x_1^{\alpha_1} x_2^{\alpha_2} \ldots x_n^{\alpha_n}$. The algebra product is defined as follows. Given $\hat{f} = \sum f_\alpha x^\alpha$, $\hat{g} = \sum g_\beta x^\beta$ we set $\hat{f}.\hat{g} = \hat{h}$ where $\hat{h} = \sum h_\gamma x^\gamma$ and $h_\gamma = \sum_{\gamma=\alpha+\beta} f_\alpha . g_\beta$.

Given real algebras V, W an <u>algebra homomorphism</u> $\phi : V \to W$ is a linear mapping with $\phi(x.y) = \phi(x).\phi(y)$ for all x, y in V.

(C1)   <u>The natural mapping $\mathcal{E}_n \to \hat{\mathcal{E}}_n$ given by $f \to \hat{f}$ is an algebra homomorphism.</u>

<u>Proof</u>   Here we keep to the notation of Chapter IV: in particular $\mathcal{M}_n, \hat{\mathcal{M}}_n$ denote the unique maximal ideals in $\mathcal{E}, \hat{\mathcal{E}}_n$. We have to show that $\widehat{(f.g)} = \hat{f}.\hat{g}$; it suffices to show that these are equal modulo an

element in $\hat{\mathcal{M}}_n^{k+1}$, for all $k \geq 0$. Choose k. By (IV.2.4) we can write $f \equiv f_k$, $g \equiv g_k$ modulo $\hat{\mathcal{M}}_n^{k+1}$ with f_k, g_k polynomials of degree $\leq k$. Then, modulo $\hat{\mathcal{M}}_n^{k+1}$, we have

$$(\widehat{f \cdot g}) \equiv (\widehat{f_k \cdot g_k}) = f_k \cdot g_k \equiv \hat{f} \cdot \hat{g}$$

bearing in mind that the Taylor series of a polynomial function is precisely that polynomial. □

Appendix D The Borel lemma

Our starting point is the explicit construction of smooth functions having very special properties.

(D1)   **There exists a smooth function** $\phi : \mathbb{R} \to \mathbb{R}$ **with the property that** $\phi(t) = t$ **for** $|t| \leq 1/2$ **and** $\phi(t) = 0$ **for** $|t| \geq 1$.

Proof   Certainly there exists a smooth function $\theta : \mathbb{R} \to \mathbb{R}$ with $0 \leq \theta(t) \leq 1$ for all t, for which $\theta(t) = 0$ precisely when $t \leq 0$, namely

$$\theta(t) = \begin{cases} 0 & \text{when } t \leq 0 \\ e^{-1/t^2} & \text{when } t > 0 \end{cases}.$$

The required function ϕ is obtained by setting

$$\phi(t) = \frac{t\theta(1-t^2)}{\theta(1-t^2) + \theta(t^2 - \frac{1}{4})}.$$

$\square$

Now we come to the Borel Lemma itself, as it was stated in Chapter IV.

(D2)   **The natural algebra homomorphism** $\mathcal{E}_n \to \hat{\mathcal{E}}_n$ **given by** $f \to \hat{f}$ **is surjective.**

Proof   We adopt the following notation. Given a sequence $x = (x_1, \ldots, x_n)$ of real numbers, and a sequence $\alpha = (\alpha_1, \ldots, \alpha_n)$ of integers ≥ 0, we write x^α as an abbreviation for $x_1^{\alpha_1} \ldots x_n^{\alpha_n}$. What we have to show is that given an element $\hat{f} = \sum \hat{f}_\alpha x^\alpha$ in $\hat{\mathcal{E}}_n$ there exists an element f in $\mathcal{E}_n$ for which $\hat{f}$ is the Taylor series of f relative to the standard co-ordinates $x_1, \ldots, x_n$ i.e. $\hat{f}_\alpha = \frac{1}{\alpha!} D^\alpha f(0)$ for all choices of α, where we write $\alpha! = \alpha_1! \ldots \alpha_n!$, $|\alpha| = \alpha_1 + \ldots + \alpha_n$ and

$$D^\alpha f = \frac{\partial^{|\alpha|} f}{\partial x_1^{\alpha_1} \ldots \partial x_n^{\alpha_n}}.$$

To this end let $\phi : \mathbb{R} \to \mathbb{R}$ be a smooth function having the properties listed in (D1) and take $\Phi : \mathbb{R}^n \to \mathbb{R}^n$ to be the smooth mapping defined by the formula $\Phi(x_1, \ldots, x_n) = (\phi(x_1), \ldots, \phi(x_n))$. We consider an arbitrary sequence (ϵ_k) of real numbers with $0 < \epsilon_k < 1$ for which $\lim_{k \to \infty} \epsilon_k = 0$, and the corresponding function $f : \mathbb{R}^n \to \mathbb{R}$ defined by $f(x) = \sum_{k=0}^{\infty} f_k(x)$ where

$$f_k(x) = \sum_{|\alpha|=k} \epsilon_k^k \hat{f}_\alpha \Phi\left(\frac{x}{\epsilon_k}\right)^\alpha.$$

Note first that f is well-defined, since for a given value of x only finitely many terms $f_k(x)$ are non-zero. Note also that $f_k(x) = \sum_{|\alpha|=k} \hat{f}_\alpha x^\alpha$ near the origin, and hence that

$$D^\alpha f_k(0) = \begin{cases} 0 & \text{if } k \neq |\alpha| \\ \alpha! \hat{f}_\alpha & \text{if } k = |\alpha| \end{cases}.$$

We claim that it is possible to choose the sequence (ϵ_k) in such a way that, on some neighbourhood of 0 in $\mathbb{R}^n$, the series $\sum_{k=0}^{\infty} D^\alpha f_k$ converges uniformly for any choice of α. It then follows from standard theorems in calculus that the sum-function f is smooth, and that

229

$$D^\alpha f(0) = \sum_{k=0}^{\infty} D^\alpha f_k(0) = \alpha! \hat{f}_\alpha$$

just as we required. By the usual comparison test it will suffice to choose (ϵ_k) in such a way that $\sum_{k=0}^{\infty} \sup_{x \in \mathbb{R}^n} |D^\alpha f_k|$ is dominated by a series (involving α) which converges for any choice of α. Clearly, we need to produce an upper bound on $D^\alpha f_k$. First of all, note that (by sheer differentiation) we have

$$D^\alpha f_k(x) = \sum_{|\beta|=k} \epsilon_k^{k-|\alpha|} \hat{f}_\beta \Psi(\alpha, \beta, k, x)$$

where

$$\Psi(\alpha, \beta, k, x) = \prod_{j=1}^{n} \frac{\partial^{\alpha_j} \phi^{\beta_j}}{\partial x_j^{\alpha_j}} \left(\frac{x_j}{\epsilon_k}\right).$$

As the function ϕ vanishes outside a compact set we have a well-defined bound

$$B(\alpha, \beta, k) = \sup_{x \in \mathbb{R}^n} |\Psi(\alpha, \beta, k, x)|$$

so setting

$$B(\alpha, k) = \sum_{|\beta|=k} \hat{f}_\beta B(\alpha, \beta, k)$$

we see that

$$\sup_{x \in \mathbb{R}^n} |D^\alpha f_k| \leq \epsilon_k^{k-|\alpha|} B(\alpha, k)$$

and hence that the series $\sum_{k=0}^{\infty} \sup |D^\alpha f_f|$ is dominated by $\sum_{k=0}^{\infty} \epsilon_k B(\alpha, k)$. All we need to do now is to choose the sequence (ϵ_k) in such a way that the latter series converges for all choices of α: and that can be done by choosing ϵ_k in such a way that $\dfrac{1}{\epsilon_k} > 2^k \sum_{|\alpha| \leq k} B(\alpha, k)$. $\square$

Appendix E   Guide to further reading

This guide is addressed largely to those who have read at least part of the present book and wish to pursue their interests, but who do not have the advantage of expert advice. Singularity theory, in common with most areas of mathematics, has an extensive technical literature, much of which is not readily available. In preparing this guide I decided only to quote those sources which I know to be available and which, in my view, a student can hope to gain from reading. I have made no attempt to compile a comprehensive list of references, on the grounds that such a list would add little to the value of the book. In particular, virtually all the mathematics in the last two chapters has its genesis in papers of J. Mather, listed for instance in [LS1].

(I)   Material Directly Related to this Book

On a general level I feel that anyone starting off in this subject should look at the classic survey paper [A1] by Arnol'd. [LS1] still provides a very good reference, and gives one a fair idea of the state of the subject at the turn of the decade. As far as more systematic texts are concerned, [L] complements the material of this volume in several respects, whilst [GG] provides a fairly formal treatment of much of the global theory.

Here are some more detailed suggestions. The beginnings of differential topology has several good expositions. First and foremost I recommend the excellent little volume [M1]. The reader who wishes to graduate to the abstract idea of a smooth manifold will find good accounts in [GP] and [BJ]. These two volumes also contain rather more material on the subject of transversality, as does the more advanced [GG]. The classification of function germs has now been taken far beyond the beginning indicated in Chapter IV, though good accounts of this fascinating area of the subject have yet to appear. Certainly the next step in this direction would be the first few pages of [A2], whilst some idea of how the subject develops can be found in [A3]. For a lucid exposition of the connexions with other areas of mathematics try [A4]. The reader who wishes to fill in the gaps in Chapter V should refer to [M], though it is not all easy going. For this one certainly needs to be familiar with the Preparation Theorem. Here I recommend Wall's "Introduction to the Preparation Theorem" in [LS1] followed by the appropriate chapter of [BL]. Further information on the classification of stable germs can only be obtained by dipping into the research literature on the subject.

(II)   Material Not Directly Related to this Book

Beyond the differentiable theory of smooth mappings, where the changes of co-ordinates are diffeomorphisms, lies the important topological theory where the changes of co-ordinates are just homeomorphisms. Some ideas concerning the local theory can be found in the latter chapters of [L], whilst an exposition of part of the global theory can be found in [GWPL]. The topological theory leans heavily on the idea of a "stratification": unfortunately, there is as yet no real introduction to the theory of stratifications, but some idea of what it is all about can be gleaned from the references just quoted. Another aspect of the topological theory is the subject matter of the classic [M2].

(III)   Applications

In Chapter IV we indicated how singularities of smooth mappings arose naturally in both differential geometry and algebraic geometry. And it is principally to these areas that one looks for applications of the theory within mathematics. As far as differential geometry is concerned I suggest looking at [P], whilst the reader who wishes to see further ahead will find solid reading in [W]. The applications of singularity theory to algebraic geometry are still at a rudimentary level, and no account of this is likely to appear for some time.

The applications of singularity theory to the physical sciences have centred around the catastrophe theory of Thom. Those with a serious interest in such matters will want to look at Thom's own book [T]. Quick introductions to the ideas can be found in [Z] and [S] while [SP] provides a comprehensive account of the state of the subject at the time of writing. For the mathematics of the matter, the account in [BL] extends the material in Chapter IV of the present volume, whilst more appears in [TZ]. I recommend [LS2] for further ideas relating singularity theory to problems in the physical sciences.

References

[A1]   Arnol'd, V. I.   Singularities of Smooth Mappings.   Russian Math. Surveys (1968) 1 - 43.   Translated from Uspehi Math. Nauk 23 (1968) 3 - 44.

[A2]   Arnol'd, V. I.   Normal Forms for Functions near Degenerate Critical Points, the Weyl Groups of A_k, D_k and E_k, and Lagrangian singularities.   Functional Anal. Appl. 6, 254-272, 1972.

[A3]   Arnol'd, V. I.   Normal Forms of Functions in Neighbourhoods of Degenerate Critical Points.   Russian Math. Surveys 29, 10-50, 1974.

[A4]   Arnol'd, V. I.   Critical Points of Smooth Functions.   Proc. Internat. Congr. Math. Vancouver 1974, 19-39.

[BL] Bröcker, T. & Lander, L.   Differential Germs and Catastrophes. London Mathematical Society Lecture Notes 17.   Cambridge University Press, Cambridge. 1975

[BJ] Bröcker, T. & Jänich, K.   Einführung in die Differentialtopologie. Springer, Berlin & New York. 1970

[GG] Golubitsky, M. & Guillemin, V.   Stable Mappings and their Singularities.   Graduate Texts in Mathematics 14. Springer, Berlin & New York, 1973.

[GP] Guillemin, V. & Pollack, A.   Differential Topology.   Prentice Hall, 1974.

[GWPL] Gibson, C. G., Wirthmüller, K., du Plessis, A. A., Looijenga, E. J. N.   Topological Stability of Smooth Mappings. Lecture Notes in Mathematics 552. Springer, Berlin & New York, 1977.

[L] Lu, Y. C.   Singularity Theory and An Introduction to Catastrophe Theory.   Springer, Berlin & New York, 1976.

[LS1] Wall, C. T. C. (Ed.)   Proceedings of Liverpool Singularities Symposium I. Lecture Notes in Mathematics 192.   Springer, Berlin and New York, 1971.

[LS2] Wall, C. T. C. (Ed.)   Proceedings of Liverpool Singularities Symposium II. Lecture Notes in Mathematics 209.   Springer, Berlin & New York, 1971.

[M] Martinet, J.   Déploiements Versels des Applications Différentiables et Classifications des Applications Stable.   Lecture Notes In Mathematics 535   Springer, Berlin & New York, 1975.

[M1] Milnor, J.   Topology - From the Differentiable Viewpoint. The University Press of Virginia. Charlottesville, 1965.

[M2] Milnor, J.   Singular Points of Complex Hypersurfaces. Annals of Maths. Studies 61 Princeton University Press. 1968

[P]   Porteous, I. R.   The Normal Singularities of a Submanifold. Journal of Differential Geometry. Vol. 5, (1971) 543-564.

[S]   Stewart, I. N.   The Seven Elementary Catastrophes. New Scientist 68, 447-454, 1975.

[SP] Stewart, I. N. & Poston, T.   Catastrophe Theory and its Applications.   Pitman.   London 1978.

[T]   Thom, R.   Structural Stability and Morphogenesis   (translated by D. H. Fowler)   Benjamin-Addison Wesley, New York. 1975

[W]   Wall, C. T. C.   Geometric Properties of Generic Differentiable Manifolds.   from Geometry and Topology III   Proceedings of Symp. at IMPA, Rio de Janeiro, July 1976.   pp. 707-774   Springer Lecture Notes no. 597.

[Z]   Zeeman, E. C.   Catastrophe Theory. Scientific American 234, 65-83, 1976.

[TZ] Trotman, D. J. A. & Zeeman, E. C.   Classification of Elementary Catastrophes of Codimension ≤ 5. Lecture Notes in Mathematics. 525 Springer.   Berlin & New York. 1976 pp. 263-327.

Index

Action of a group 61
Algebra 226
Algebra product 226
    homomorphism 226
Algebraic set 222

Bifurcation set 44
Binary cubic form 65, 66
Boardman symbol 180, 181
    submanifold 186
Borel Lemma 101, 228
$\mathscr{C}$-equivalence 144
Chain Rule 9, 21, 23
Change of basis 72
    of co-ordinates 72
    of parameter 90
Chart 13
Codimension of function germ 98
    of map germ 152
    of orbit 77
    of submanifold 15
Co-ordinates, local 13
Corank of function germ 125
Critical point 29, 40
Critical point, degenerate 59

Critical value 40
Cubic curves 68-70
Cuspoids 130
Curve 25
Deformation of germ 140
    induced 163
    linear 56
    transversal 160
    universal 164
Dense 52
Descendant 110
Determinacy 116, 191
Determinacy, finite 120, 191
Diffeomorphism 8, 12
Differential 8, 20
Double point 130

Equivalence of deformations 163
    of germs 139
    of maps 16, 205
    of unfoldings 89, 142

Fold curve 46
Folded handkerchief mapping 55
Flow line 26

Fundamental neighbourhood	51
General linear group	63
Generic	187
Germ, equivalence of	35
immersive	35
invertible	34
rank of	35
representative of	34
singular	35
source of	34
submersive	35
target of	34
Graph	21
Hadamard Lemma	100
Hessian	59, 67
Homogeneity	62
Homomorphism, induced	112
Infinitesimally stable point	78
Inverse function theorem	10
Isomorphism of ideals	149
Isotropy subgroup	80
Jacobian extension	180
ideal	97
module	151
Jet extension	37
space	37
Jets	37
Jets, equivalence of	62
of germ	37
$\mathcal{K}$-equivalence	143, 144
Lie group	73
Linear systems	71
Local Existence and Uniqueness Theorem	28
flow	28
Manifold	12
Metastable range	212
Morphism of deformation	164
Multiplicity	130
Nakayama Lemma	102
Normal bundle space	133
Null set	215
Orbit	61
Parametrization	12
Pencil	71
Product structure	80
Quadratic form	64
Quadratic form, index of	65
rank of	65
semi-index of	65

R - equivalence	94
Rational mapping	223
Regular point	224
value	40
Sard's Theorem	49, 217
Semialgebraic set	222
Semi-direct product	150
Singular point	44
Singular point,	
non-degenerate	123
of type A_k	130
Singularity set	44, 54
Singularity set,	
first order	55
higher order	177
Slice	79
Smooth mapping	8, 12
Splitting Lemma	125
Stable germ	142
map	206
point	77
Submanifold	15
Submersion	40
Tangent bundle	24
bundle space	22
mapping	22
space	16
Tarski-Seidenberg Theorem	224

Transversality Lemma, Basic	49
Transversality Theorem	47
Transversality Theorem,	
elementary	51
of Thom	53
Transverse	39
Transverse intersections	38
Triple point	130
Umbilic bracelet	68
Unfolding	81
Unfolding of germ	141
induced	89
minimal	81
morphism of	90
transverse	81
universal	90
versal	90
Vector field	24
Vector field, equivalence of	30
time dependent	32
time independent	33
Whitney cusp mapping	44
Whitney umbrella	87, 213

RAYMOND H. FOGLER LIBRARY

QA
614.58
G53

APR 9 1981